Kudos for this book fro

♦ Too often mathematics is displ ... only in furthering itself. Mr. Lauba ... retained his sense of delight with the potentials of mathematics and he shares some of his fun applications in A Mathematical Medley. Uses in the real world of physics are sprinkled throughout, interspersed with some twists and turns in computations that relatively elementary mathematics can illuminate. His sense of humor is evident in the goodies filling the interstices of the book. This is an altogether delightful collection of applications of mathematics.
Walter W. Whitman, Prof. Emeritus of Geophysics, and sometime Prof. of Mathematics, Colorado School of Mines

♦ Professor Laubach has shared his passion for mathematics in this fascinating collection of problems. Teachers and students alike will find food for thought; both entertaining and challenging. The birthday problem is a winner!
Elaine Fitzgerald, Mathematics Teacher, Golden High School

♦ If you love math you will be challenged. If you love success stories you will be delighted and inspired. Winton Laubach's words and math problems/riddles are destined to touch thousands of lives. Just buy the book and try to figure out the answers.
Joe Sabah, President of the Colorado Independent Publishers Association and author of How to Get On Radio Talk Shows All Across America Without Leaving Your Home or Office.

♦ A love of mathematics has inspired Professor Laubach to pass it on in this most interesting medley of imaginative topics.
Ross R. Middlemiss, Retired Professor of Mathematics, Washington University. St. Louis, Missouri, and author of Differential and Integral Calculus; College Algebra; and Analytical Geometry.

This book is dedicated to my wife, Janet,
and the family she gave me.

A MATHEMATICAL MEDLEY

Cover Design by Shadow Canyon Graphics
(303) 278-0949

Printing by Millennia Graphics
(719) 638-0991

A MATHEMATICAL MEDLEY

Gleanings From The Globe And Beyond

Winton Laubach

Aftermath Publishing, Golden, Colorado

A MATHEMATICAL MEDLEY

Gleanings From The Globe And Beyond

by Winton Laubach

Published by:
Aftermath Publishing
1490 Rogers Street
Golden, Colorado 80401

Printed in the United States of America

Publisher's Cataloging-in-Publication
(Provided by Quality Books, Inc.)

Laubach, Winton.
A mathematical medley : gleanings from the globe and beyond / Winton Laubach. -- 1st ed.
p. cm.
Includes index.
LCCN: 99-94703
ISBN: 0-9670087-0-0

1. Mathematical recreations. 2. Physics--Problems, exercises, etc. I. Title.

QA95.L38 1999 793.7'4
QBI99-423

FOREWORD

It was with some trepidation that I, a Liberal Arts graduate, signed my first contract in 1955 to teach mathematics at the Colorado School of Mineral Engineering in Golden. My fears were not totally unfounded, but there were a few faculty members who welcomed me and made me feel at home. Winton Laubach was one of them.

As time passed I continued to consider him a valued friend and colleague while also coming to recognize and respect his knowledge of many fields only loosely associated with the standard mathematics curriculum. During my teaching career, I found some of his favorite problems (e.g., the Snowplow Problem) becoming favorites of mine as well. Those I then shamelessly appropriated for use in many of my classes.

At meetings of the Mathematics Club one could always count on Winton to come up with a provocative problem, question or idea. Ranging from plane geometry and spherical trigonometry to statistics, number theory and celestial mechanics, the diversity of his interests are now evinced in the pages of this book.

There is much, much more that I would like to tell the reader about the author, but I realize that Forewords are boring (if they are even read), so I'll close now with the warning to the reader that he (or she) should be prepared to exercise his (or her) brain when reading A Mathematical Medley. However, I assure you that the pleasure derived therefrom will prove to be well worth the effort expended.

Dr. D.C.B. Marsh, Professor Emeritus, Colo. School of Mines and Past National President, American Cryptogram Assn.

ABOUT THE AUTHOR

The author did the most important part of his growing up on a small farm near small Benton, Pennsylvania. Possessing a lazy streak, nothing he accomplished during his formative years suggested that he showed promise. And nothing that transpired thereafter appreciably altered that outlook. There were moments of glory, however. His eighth grade class entered an essay contest sponsored by the Women's Christian Temperance Union. At that time the author was quite sympathetic to their cause and with that motivation won the three dollar first prize. It didn't end there. When the essay was entered in the county competition, he was again awarded first prize. This time the monetary award was two dollars. The W C T U later came to believe that these awards had been a waste of money.

After graduating from nearby Bloomsburg State Teachers College, he subsequently spent two and a half years teaching at New Milford, Pennsylvania, High School. This was followed by graduate studies in mathematics at Columbia University. While there, he was privileged to be in attendance when the Winston Churchills were greeted by the Dwight Eisenhowers. At that time, Eisenhower was president of Columbia University and Churchill had come to visit following his "Iron Curtain" speech.

The author's college teaching career started at New York state's Sampson Naval Base. It had been transformed into a temporary college for GIs returning from the second world war. This was followed by a five year stint at Penn State University and then eighteen years at the Colorado School of Mines. At this point, three events contributed an overall improvement to the author's well-being. He lost his eyesight, lost his job, and found a wife.

In addition to his wife and mathematics, the author's interests are his friends, cryptograms, crossword puzzles, Dixieland jazz and harmonizing with any singers who can carry a tune.

PREFACE

On many occasions in my career as a teacher of mathematics, I wished that I could share with the world the surprises and thrills from newly discovered revelations. The yearning never waned, and that is the reason for this book.

The topics of this book were chosen to satisfy those with expertise, as well as those with just a peripheral interest in science. These choices resulted in topics rarely or never included in a standard mathematics or physics curriculum. The subjects are frequently unconventional, often requiring unsophisticated mathematics, and are intended to interest the student, mathematician and elementary physicist. Also, the book is designed to appeal to that segment of the general public which enjoys knowledge for knowledge's sake. Several of the chapters should be interesting and comprehensible to those with minimal mathematical training. The table of contents identifies those chapters with an asterisk. Only five chapters involve elementary calculus. The quintet of heavenly body chapters appearing near the end of the book may be the most difficult to comprehend, but the tenacious reader may find them to be the most satisfying.

The book is by necessity short, since it contains nearly all that I know. All humor that does not overly detract from the gravity of the situation being addressed, has been allowed to remain. There is even a little something for those with short attention spans. Wherever a chapter ends with a blank page, there is an illuminating and/or entertaining morsel that may be directly or vaguely related to mathematics.

I am blind, so this book could not exist without extensive help. Technical assistance was provided by Bruce Law. On various occasions, stepson Andrew Middlemiss rescued the endeavor with his skillful manipulation of the software. The programming required to obtain the tabular values in Chapters Six and Eleven was the work of my brother, John Laubach. Friends Donald Rabb, Joseph Davis and Fred Carpenter provided information and motivation.

Friend Norma Frey wore many hats. She was the able mathematician who edited this work and caught me whenever I stumbled. Computer generation of diagrams, equations and symbols were beyond my capability. There again Norma was the savior. The concept for the cover was her contribution as was the book's subtitle. Her commitment and enthusiasm were the catalysts that kept me focused.

To top it off, my wonderful wife, Janet, constantly rescued me from bungling performances on the computer and was ever ready with her valuable eyesight and insight. Though lacking a mathematical background, first drafts of diagrams described to her by a blind husband were superbly constructed. She persevered through the long list of publishing tasks with consistently superior efficiency and good humor. Again she demonstrated how fortunate I am to have wound up with her instead of a guide dog.

Winton Laubach
Golden, Colorado

TABLE OF CONTENTS

TABLE OF CONTENTS

TABLE OF CONTENTS

*Requires Little Mathematical Training

Lo the poor mathematician,

ever discouraged from drinking and deriving.

CHAPTER 1

THE COW THAT WALKED THE INVOLUTE

My first memorable thrill from mathematical creativity resulted from a conversation in one of my classes at a northeastern Pennsylvania high school. The farmer father of one of my students wanted to know how far a cow would walk if, tethered to a stake one inch in diameter by a fine 100 foot wire, she kept the wire taut and walked until she wound up to the stake. As often happens with new teachers, my mettle was being tested by the father. Fresh out of teacher's college, my exposure to calculus had been too rudimentary to be of any assistance. I contemplated estimating the average distance walked during one circuit of the stake and multiplying by the number of circuits walked, but this would satisfy neither him nor me. It was a situation that required an inspiration. In time it came, and the following discourse is the result.

Before we can consider the problem of the cow, it will be helpful to refer to the regular polygon shown in the diagram on page 2.

This regular polygon of n sides is inscribed in a circle of diameter d. We will tightly wrap a fine wire of length w around the exterior of the polygon. For convenience, we will choose a length that is a multiple of the length of a side of the polygon. It will become apparent later that this restriction will produce no loss of generality. We choose the following variables:

p = the perimeter of the polygon
n = the number of sides of the polygon
s = the length of a side of the polygon
w = the length of the wire
d = the diameter of the stake

S = the length of the approximation path as will soon be defined

r = the ratio between the length of the wire and the length of a side of the polygon

We see then that $p = ns$ and $w = rs$.

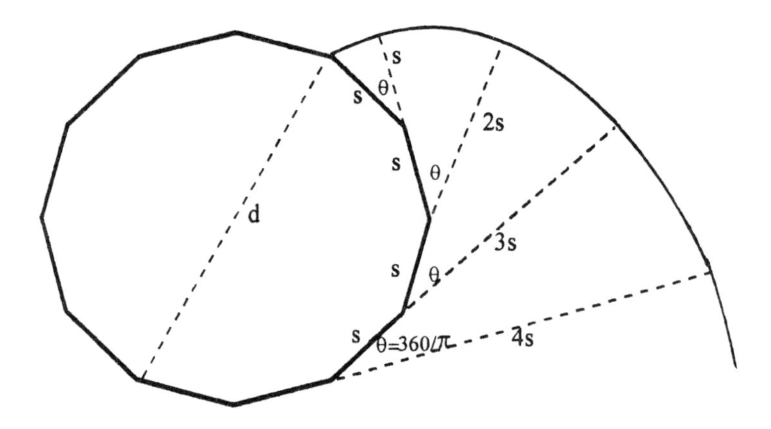

Now to achieve our objective, we wrap the wire about the polygon, being careful to have the end of the wire at a vertex of the polygon. As a consequence of the definition of r, r represents the number of sides that will be covered by the wrapped wire. Keeping the wire taut, unwind it until it is in line with a side of the polygon. As the wire is unwound, its initial end describes a sequence of arcs of circles with radii whose lengths are $s, 2s, 3s, \ldots, rs$, respectively. Plane geometry teaches us that the central angle for each arc is $360°/n$, hence each arc is $1/n$ times its respective circumference.

As the wire is unwrapped, the end describes a path whose total length is

$$\frac{2\pi s}{n}(1+2+3+\cdots+r)$$

Using the function $\frac{k(k+1)}{2}$ for the sum of the first k positive integers, and noticing $k = r$ in our example, we see that

$$S = \frac{\pi sr(r+1)}{n}.$$

Recalling that $w = rs$ and $p = ns$, we obtain

$$S = \frac{\pi sw}{p\left(\frac{w}{s}\right)+1}$$

$$= \frac{\pi w}{p}(w+s).$$

Finally we let n, the number of sides of the polygon, increase without limit. As a consequence, the shape of the polygon will approach the shape of its circumscribed circle, p will approach the circumference of the circle, πd, s approaches 0, and the path composed of the set of r arcs of circles will approach the path of the cow. Since p approaches πd and s approaches 0, as n increases without limit, the approximation curve length

$$S = \frac{\pi w}{p}(w+s) \text{ approaches } \frac{w^2}{d}.$$

These concepts provide the length of path, P, as given below.

$$P = \frac{w^2}{d}$$

The curve described is called an involute of a circle.

For the farmer's problem, $w = 100$ feet and $d = 1/12$ foot. Of course, a tether to an untrained cow doesn't quite conform to our idealized conditions, but we will apply the formula anyway. The answer is 120,000 feet or almost 23 miles.

An analogous procedure will provide a formula for the area swept out by the wire as it winds (or unwinds) about the stake. Simply replace the series of arc lengths by the corresponding series of areas of sectors of circles,

$$\pi/n[s^2 + (2s)^2 + (3s)^2 + \cdot \cdot \cdot + (rs)^2].$$

Whereas previously we had a sum of integers from 1 to r, we now have a sum of squares of integers from 1 to r. Using the expression $\frac{k(k+1)(2k+1)}{6}$ for the sum of the squares of the first k integers, and applying the substitutions used in the first problem, the sum of areas of the r sectors can be transformed into the expression $\frac{\pi}{6p}w(w+s)(2w+s)$. Again, as n increases without limit, p approaches πd, s approaches 0, and hence the sum of the areas of the sectors approaches $\frac{w^3}{3d}$. So the area swept out by the rotating wire is given by the formula below.

$$A = \frac{w^3}{3d}$$

Of course, when more than one circuit of the stake occurs, there will be duplication of areas. Notice that the expression representing P is the derivative of the expression representing A. This is fascinating, maybe not surprising, but hardly predictable.

Now let us compare the initial result obtained above with that obtained by the approximation method previously mentioned. Choosing the estimates of the average length of the circuit about the stake to be the circumference of the circle with radius $w/2$ and realizing that the number of circuits would be the length of the wire divided by the circumference of the stake, the approximation turns out to be $2\pi\left(\frac{w}{2}\right)\left(\frac{w}{\pi d}\right)$.

Coincidentally this equates to the newly familiar $\frac{w^2}{d}$. So some readers may think that this has been a foolish exercise in redundancy. I hope not.

A Morsel (food for thought)

"A mathematician is a blind man in a dark room looking for a black cat that isn't there"

Charles Darwin

CHAPTER 2

*PIANOS ARE TACHOMETERS

"I'd like to know how fast that is spinning." This wonder about a gyroscope came from my stepson at a family gathering. It was a question worth considering, and it provided the opportunity to try an experiment that had previously occurred to me.

The plan was to blow through a straw forcing a blast of air through the rotating evenly spaced spokes of the gyroscope. It was assumed that the passage of each spoke through the moving column of air would create a vibration in the column. The regularity of these vibrations should then produce a musical note. This note would be recorded on tape and the frequency of the note should enable us to determine the rotational speed of the gyroscope. But here we were temporarily stymied. The frequency of piano notes was not readily available. Was there a method to determine piano frequencies that would free us from relying on a handbook or calculator? In time, trial and error led to the discovery of the simple quadratic expression on the next page. It gives surprisingly accurate values for the frequencies. For the octave between A-440 and A-880, the values provided by the formula are as follows: A# is 1 vibration per second too large, F# is 1 vibration per second too small, G is 2 vibrations per second too small and G# is 4 vibrations per second too small. The remaining frequencies are all exact to the nearest integer.

In the following quadratic expression, n represents the number of white and black keys, counting to the right from A-440 to the note in question. A-440 is four octaves above the lowest note on a standard piano. The A notes are emitted by the white keys which immediately precede the third black key in each group of three black keys.

To obtain a note's frequency, calculate the value of the expression displayed below. As stated previously, the variable n which occurs there and later, represents the number of keys that the note of interest is to the right of A-440.

$$n(n + 24) + 442$$

For example, let us determine the frequency of the C# which is in the first octave above A-440. The keys from A-440 are A#, B, C then C#. So, for our example, n is equal to 4. Therefore the value of the frequency expression is $4 \times 28 + 442 = 554$, which is the frequency of our C#. For notes above the octave we are discussing, multiply by 2 for each octave difference. For octaves below, divide by 2 for each octave difference.

With these results, it was possible to finish our experiment. The recorded note from the gyroscope turned out to be the C, one octave above our base octave. Since C is 3 keys above A-440, the frequency of the note recorded was double the frequency of the base octave C, for a total of twice the value of $3 \times 27 + 442$. Hence the grand total was 1046 vibrations per second. Since the brain, through the medium of a piano, will identify only the note whose frequency is closest to the frequency recorded, we knew only that the frequency was closer to that of C than to the frequencies of B or C#. After computing the frequencies of the latter two notes, we learned that the rate at which the spokes passed the straw was between 988 and 1108 spokes per second. But to determine the revolutions per second for the gyroscope, we needed to divide the preceding numbers by the number of spokes that passed the straw during one revolution. This number was 6, so for the short time interval of the puff of air, the gyroscope was rotating at a rate between 165 and 185 revolutions per second.

This revelation about the revolution made our day.

As already mentioned, the quadratic expression that we have been using provides a very good approximation to the piano frequencies in a particular octave. For those who find scientific calculators user friendly, the following function of k provides the theoretically correct frequency for every note on a standard piano. The variable k represents the key number, starting the count at the left end of the piano.

$$27.5\left(2^{\frac{(k-1)}{12}}\right)$$

For example, suppose we want to determine the frequency of the top note on the piano. In this case the value of k is 88, the exponent of 2 is 7.25 and the resulting frequency is 4186. By comparison, if we apply the original procedure to approximate the frequency of this same top note, C, we obtain 4184.

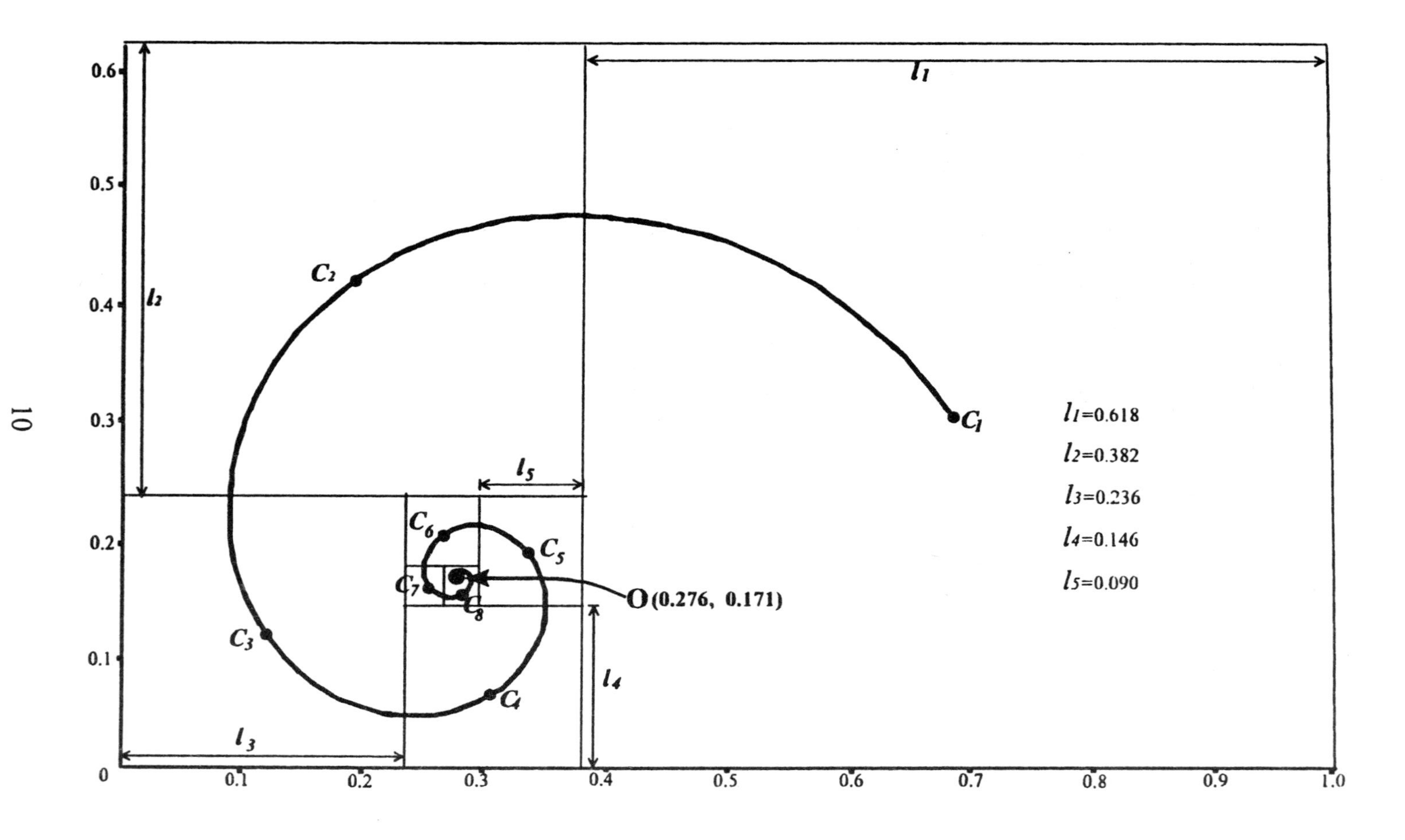

l_1=0.618
l_2=0.382
l_3=0.236
l_4=0.146
l_5=0.090
O (0.276, 0.171)
C_1
C_2
C_3
C_4
C_5
C_6
C_7
C_8
l_1
l_2
l_3
l_4
l_5
0
0.1
0.2
0.3
0.4
0.5
0.6
0.7
0.8
0.9
1.0

CHAPTER 3

GILDING THE GOLDEN RECTANGLE

Once upon a time I was killing time. A friend had recently sent a clipping commenting on an interesting property of a certain rectangle which mathematicians call a golden rectangle. It occurred to me that this was a convenient topic on which to focus my idle curiosity. It turned out to be a fortunate choice, as the next two chapters will reveal. These chapters will evolve entirely from principles that are reflected in the diagram on the facing page.

The large rectangle displayed is a rectangle whose width has the measure r and whose length has the measure 1. We will determine what the value of r must be in order that the ratio of the width to the length is the same as the ratio of the length to the width plus length. This relation between width and length determines that our rectangle will be a golden rectangle. Algebraically this means that dimension r must satisfy the equation $\frac{r}{1} = \frac{1}{r+1}$. This results in the equivalent equation $r^2 + r - 1 = 0$, for which the only positive solution is $\frac{\sqrt{5}-1}{2} = 0.618033989$. From the preceding equation the following useful relations (labeled Rel 1., Rel 2., Rel 3.) may be inferred.

Rel 1. $r^2 = 1 - r$

Rel 2. $r = 1 - r^2$

Rel 3. $\frac{1}{r} = r + 1$

We are now prepared to analyze the included golden rectangle, as well as its internal configuration. The diagram reflects the crucial value r which initially measures the large rectangle's width. This rectangle is partitioned into the large square, with its center at C_1 at the right and the accompany-ing rectangle along its left side. This construction requires that the width of the latter rectangle be $1 - r$. Rel 1. informed us that $1 - r = r^2$. Hence the width/length ratio of the latter internal rectangle is $\frac{r^2}{r} = r$, implying that this rectangle is also a golden rectangle. This second rectangle is in turn partitioned into a square at the top with center at C_2, its accompanying golden rectangle at the bottom. Then this third rectangle is partitioned into a square at the left with its center at C_3 and its golden rectangle to the right. This process is repeated ad infinitum, rotating in a counterclock-wise direction, partitioning each ensuing rectangle into a square and additional golden rectangle.

The points C_1 through C_8 indicate the centers of the first eight squares in the never ending sequence of nested diminishing squares. The dimensions of the first five squares are given in the diagram and indicate that the dimension of each succeeding square is r times the dimension of the square it follows. As we continually increase the number of squares in the sequence, their centers must approach ever closer to a limiting point. This is the point O displayed in the diagram. Determining the coordinates x_0 and y_0 of this point O proves to be an interesting exercise which will now be explored. Notice that square number 5 and its accompanying rectangle are aligned in the same position as are square number 1 and its accompanying rectangle – that is, square to the right and rectangle to the left. This means that the configuration from square 1 through square 5 is the same as the configuration from square 5 through square 9, which is in turn the same as the configuration from square 9 through square 13, etc. The left border of each additional square in this sequence of squares 5, 9, 13, . . . , approaches ever closer to the point O as the number of squares in the sequence increases without limit. Consequently the difference between x_0 and the x-coordinate of the right side of square 5, r^2, is the sum of the infinite series created by adding the dimensions of squares 5, 9, 13, . . . , ad infinitum. This series is the geometric series $r^5 + r^9 + r^{13} + \cdots$ with a common ratio r^4. The sum of this infinite series is known to be $\frac{r^5}{1-r^4}$. Though x_0 could be computed at this stage we can apply Rel 3. to find x_0 and discover yet another interesting relationship. In $\frac{r^5}{1-r^4}$ we repeatedly divide numerator and denominator by r, apply Rel 3. and obtain:

$$\frac{r^5}{1-r^4} = \frac{r^4}{r+1-r^3} = \frac{r^3}{r+2-r^2} = \frac{r^2}{r+3} = \frac{r}{3r+4} = \frac{1}{4r+7}.$$

x_0 is then obtained by subtracting this result from r^2, the x-coordinate of the right side of square 5. We won't do it here, but with the aid of Rel 3. it can be shown that the result is

$$x_0 = \frac{1}{r+3}.$$

Likewise, the configuration of nested squares from square 4 to square 8 is the same as the configuration of squares from square 8 to square 12, etc. As a result the point O is approached from below by the top border of the top square in the encroaching sequence of squares 4, 8, 12, ... Thus y_0 is the sum of the infinite series $r^4 + r^8 + r^{12} + \cdots$. The sum of this series is $\frac{r^4}{1-r^4}$. Using the procedure applied when determining x_0, it is possible to reduce this result to find $y_0 = \frac{1}{3r+4}$.

When we get down to the nitty-gritty of the arithmetic we learn that the coordinates of point O are (0.276, 0.171).

If we again apply Rel 3. we can show that $r \times x_0 = y_0$. Using the techniques of analytic geometry, it can be shown that a line segment connecting points C_1 and C_3 is perpendicular to a line segment connecting points C_2 and C_4. And there's more! The center of each odd numbered square lies on the segment connecting C_1 and C_3 while the center of each even numbered square lies on the segment connecting C_2 and C_4. All of which leads us to the obvious corollary, the *coup de grace*. These two preceding segments intersect at point O.

If we again apply the tools of analytic geometry, we learn that the distance OC_2 divided by the distance OC_1 is equal to r. In general the distance OC_n is equal to r times the distance OC_{n-1}:

$$OC_n = r \times OC_{n-1}.$$

One question remains. Is it possible to determine the polar coordinate equation of the spiral passing through the centers of all the squares? It turns out that it is and it follows.

With the polar coordinate origin at point O, we obtain the spiral's polar equation $\rho = f(\theta)$ as follows: Segments OC_{n+1} and OC_n are known to be perpendicular and $OC_{n+1}/OC_n = r$. Hence the $f(\theta)$ must be chosen such that $f(\theta + \frac{\pi}{2}) / f(\theta) = r$, ($\theta$ radians). This condition suggests $\rho = kr^{\frac{2\theta}{\pi}}$. Known rectangular coordinates of points O and C_1 imply that the polar coordinates of C_1 are $(\frac{r}{\sqrt{2}}, Arc\cot 3)$.

Substituting these coordinates in the previously suggested polar equation, solving for k and simplifying k provides this spiral equation:

$$\rho = 0.5\sqrt{2}\, r^{-\frac{2\,Arc\tan 3}{\pi}}\, r^{\frac{2\theta}{\pi}} = 0.4823 r^{\frac{2\theta}{\pi}}$$

CHAPTER 4

THE FIBONACCI-GOLDEN RATIO CO

We are doubly blessed. The ratio r, the width/length ratio of the "golden rectangle", served well as the basis for the preceding chapter. It also provides the foundation on which we will build the current chapter.

As was previously stressed, this ratio r became the positive root of the equation, $r^2 + r - 1 = 0$.

As a result $r = \dfrac{\sqrt{5}-1}{2} = 0.618033989$. It is vital to notice that this number is less than 1.

Occasionally, little things mean a lot. This is especially true for the simple relation $r^2 = 1 - r$ which follows from the equation of the preceding paragraph. Repeated application of this relation informs us that:

$$r^2 = -r + 1$$
$$r^3 = r(1 - r) = r - (1 - r) = 2r - 1$$
$$r^4 = r(2r - 1) = 2(1 - r) - r = -3r + 2$$
$$r^5 = r(-3r + 2) = -3(1 - r) + 2r = 5r - 3$$
$$r^6 = r(5r - 3) = 5(1 - r) - 3r = -8r + 5$$

We now temporarily digress to examine a much explored sequence of numbers. Consider the sequence defined as follows: Initiate the sequence with two successive 1's. Then each succeeding term is the sum of the preceding two terms. The result is 1, 1, 2, 3, 5, 8, 13, 21, 34, ... This is the well known Fibonacci Sequence.

Returning to our original sequence, the sequence of powers of r, it appears to contain two internal subsets such that in each subset the absolute values of the elements are Fibonacci numbers. Both the sets 1, 2, 3, 5, 8 (the absolute values of coefficients of r) and 1, 1, 2, 3, 5 (the absolute values of the constants) conform to the prescribed pattern. Mathematical induction allows us to prove that the suggested pattern of Fibonacci numbers does indeed persist ad infinitum.

It is now advantageous to represent the value of the nth number in the Fibonacci sequence with the symbol $F(n)$. As a result:

$$r^2 = -F(2)r + F(1)$$
$$r^3 = F(3)r - F(2)$$
$$r^4 = -F(4)r + F(3)$$
$$r^5 = F(5)r - F(4)$$
$$r^6 = -F(6)r + F(5)$$

Generalization produces: $r^n = (-1)^{n+1}[F(n)r - F(n-1)]$

From the preceding equation we learn that

$$\frac{F(n-1)}{F(n)} = r + (-1)^n \frac{r}{F(n)}.$$

Recalling that r < 1, we see that, as *n increases without limit,* $\frac{F(n-1)}{F(n)}$ *rapidly approaches r*. This knowledge that the ratio $\frac{F(n-1)}{F(n)}$ has the limit r will supply the foundation for that which remains. (For the remaining discussion we will practically abandon the ratio r.)

From this point on it will be convenient to recall the golden rectangle, but we will convert from the ratio $r = \frac{\text{width}}{\text{length}}$ to the ratio $R = \frac{\text{length}}{\text{width}}$. Consequently the golden rectangle is now defined as the rectangle with width 1 and length R such that $\frac{R}{1} = \frac{R+1}{R}$. This implies that $R^2 = R + 1$ or $R^2 - R - 1 = 0$. The positive root of this equation, 1.618033989, will be the number R of the remaining discussion. Notice that $R = \frac{1}{r} = r + 1$. This R is the conventional choice for the "golden ratio". Since we have learned that $\frac{F(n-1)}{F(n)}$ approaches r as n increases, we now know that $\frac{F(n)}{F(n-1)}$ must approach $\frac{1}{r} = R$ as n increases without limit. Thus we have this fascinating result:
The Fibonacci sequence is a quasi-geometric sequence with R masquerading as the "common ratio".

We now use the equation, $R^2 = R + 1$, to show the connection between powers of R and the Fibonacci sequence.

$R^2 = R + 1$
$R^3 = R(R + 1) = R^2 + R = R + 1 + R = 2R + 1$
$R^4 = R(2R + 1) = 2(R + 1) + R = 3R + 2$
$R^5 = R(3R + 2) = 3(R + 1) + 2R = 5R + 3$
$R^6 = R(5R + 3) = 5(R + 1) + 3R = 8R + 5$

Again the arrangement of Fibonacci numbers in these expressions implies that perhaps $R^n = F(n)R + F(n-1)$ for all positive integral values of n. Not suprisingly, mathematical induction verifies that this is indeed the case.

The preceding discussion guarantees that every positive integral power of R reduces to a linear function of R. The equation defining R implies that $R^{-1} = R - 1$. Consequently $R^{-n} = (R - 1)^n$. Since the expansion of this binomial produces a polynomial in R and every term of the polynomial is a linear function of R, R^{-n} must also be a linear function of R. This relation combined with the previous knowledge that positive integral powers of R reduce to linear functions of R, implies that: *Every integral power of R reduces to a linear function of R.*

The table on the next page illustrates some of the information just discussed. Only a small sampling of the Fibonacci numbers is represented. A more extended list appears at the end of the chapter. Column 2 lists the relevant Fibonacci numbers. Column 3 displays the ratio between the Fibonacci number there listed and the preceding Fibonacci number. Notice that this ratio, $\frac{F(n)}{F(n-1)}$, is less than R when n is even, and greater than R when n is odd. Column 3 also indicates that for n equal to 25, the ratio $\frac{F(n)}{F(n-1)}$ and R are already identical through the first ten significant digits. . A result obtained earlier implies that, as n increases, the difference between R and $\frac{F(n)}{F(n-1)}$ approaches zero. Column 4 displays the specific relation between Fibonacci numbers and the values of the positive integral powers of R. Remember that $R = 1.618033989$.

1	2	3	4
n	$F(n)$	$\frac{F(n)}{F(n-1)}$	R^n
1	1		R
2	1	1	R+1
3	2	2	2R+1
4	3	1.5	3R+2
5	5	1.667	5R+3
9	34	1.61905	34R+21
10	55	1.61765	55R+34
14	377	1.618026	377R+233
15	610	1.618037	610R+377
19	4181	1.618030341	F(19)R+F(18)
20	6765	1.618033963	F(20)R+F(19)
24	46,368	1.618033988	F(24)R+F(23)
25	75,025	1.618033989	F(25)R+F(24)

It is interesting to note how closely for large values of n, the Fibonacci sequence approximates a geometric sequence. If $F(20)$, 6765, is multiplied by R^{20}, the result when rounded off to the nearest integer, is 102,334,155. This is precisely the value of $F(40)$ as given in the list of the first forty Fibonacci numbers at the end of this chapter.

The following theorems were motivated by inspection of that group of Fibonacci numbers just referred to.

Theorems:

1. $F(n)$ is a multiple of 2 if and only if n is a multiple of 3.
2. $F(n)$ is a multiple of 3 if and only if n is a multiple of 4.
3. $F(n)$ is a multiple of 4 if and only if n is a multiple of 12.
4. $F(n)$ is a multiple of 5 if and only if n is a multiple of 5.
5. When combined, Theorems 1 and 4 imply that $F(n)$ is a multiple of 10 if and only if n is a multiple of 15.

Theorems for other specific integers may also be developed, but the following theorem may suffice.

Theorem 6:

$F(m)$ is a multiple of $F(n)$ if and only if m is a multiple of n.
Specifically,
$F(2n) = F(n)F(n-1) + F(n+1)\,F(n) = F(n)\ [F(n-1) + F(n+1)]$
$F(3n) = F(n)\,F(2n-1) + F(n+1)\,F(2n)$
$F(4n) = F(n)\,F(3n-1) + F(n+1)\,F(3n)$, and so forth.

When performing a cursory search for primes among the $F(n)$'s, searching ceased after the investigation of $F(23)$. Within this group, the only prime $F(n)$ for which n is not also a prime, is $F(4) = 3$. The only prime value of n in this group for which $F(n)$ is not also prime is 19 since $F(19) = 4181 = 37\times113$.

For those who might want to add a group of consecutive terms of the Fibonacci series, the following formula may be useful:

$$\sum_{n=s}^{t} F(n) = 2F(t) + F(t-1) - F(s+1) = F(t+2) - F(s+1).$$

A far more astounding relation was supplied by a friend who was aware of my interest:
The sum of any ten consecutive Fibonacci numbers is always equal to eleven times the seventh term.

An interesting relative of the golden ratio family of facts is the following simple continued fraction .

$$\cfrac{1}{1+\cfrac{1}{1+\cfrac{1}{1+\cfrac{1}{1+\cfrac{1}{1+\cdots}}}}}$$

Notice that the continued fraction which is initiated by the 2nd term in the denominator of the top fraction replicates the initial continued fraction itself. In other words, if we let the symbol x represent the entire continued fraction, it follows that $x = \dfrac{1}{1+x}$.
Since this transforms into $x^2 + x - 1 = 0$, the original equation which defined r, it follows that the continued fraction must converge to $r = \dfrac{1}{R}$. Evaluations at successive positions of the continued fraction produce approximations that are successive ratios of consecutive pairs of the Fibonacci numbers partially listed on the next page.

(1) 1
(2) 1
(3) 2
(4) 3
(5) 5
(6) 8
(7) 13
(8) 21
(9) 34
(10) 55
(11) 89
(12) 144
(13) 233
(14) 377
(15) 610
(16) 987
(17) 1,597
(18) 2,584
(19) 4,181
(20) 6,765
(21) 10,946
(22) 17,711
(23) 28,657
(24) 46,368
(25) 75,025
(26) 121,393
(27) 196,418
(28) 317,811
(29) 514,229
(30) 832,040
(31) 1,346,269
(32) 2,178,309
(33) 3,524,578
(34) 5,702,8 87
(35) 9,227,465
(36) 14,930,352
(37) 24,157,817
(38) 39,088,169
(39) 63,245,986
(40) 102,334,155

CHAPTER 5

At The Drop Of A Needle

My interest in this problem originated while eavesdropping on colleagues who were discussing it. The problem possessed a happy combination. It was intriguing and provided a convoluted procedure for estimating the value of π.

PROBLEM: A needle of length N is randomly thrown onto a hardwood floor whose boards are all of width W. If the length of the needle is less than the width of the boards, determine the probability that, as the needle comes to rest, it will intersect a joint between boards.

Letting y be the distance of the needle's center from the nearer joint, and θ be the acute angle formed by the needle and joints, we see that the needle will intersect a joint if and only if $y \leq \frac{N}{2}\sin\theta$. We note that y and θ are independent random variables. Their domains are $0 < y < \frac{W}{2}$ and $0 < \theta < \frac{\pi}{2}$. Given these domains, we see that the set of all such points would create the rectangle below.

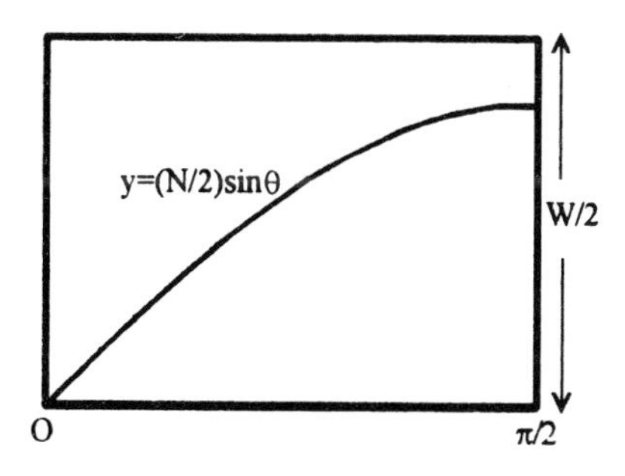

We see that the inequality is satisfied if and only if the y-coordinate of a random point of the rectangle is bounded above by the sine curve. The probability of this is equal to the area under the sine curve divided by the area of the rectangle. The area under the curve is:

$$\int_0^{\pi/2} \frac{N}{2} \sin\theta d\theta = \frac{N}{2}$$

Hence, the probability is

$$P = \frac{\frac{N}{2}}{\frac{\pi W}{4}} = \frac{2N}{\pi W}$$

As the number of trials increases the ratio of the number of needles intersecting a joint to the number of needles thrown, approaches the probability P. Letting j represent the number of needles intersecting a joint and t represent the number thrown, we then know that $\frac{j}{t}$ approaches $\frac{2N}{\pi W}$. This implies that $\frac{2Nt}{Wj}$ approaches the value π.

It's an experiment worth enjoying. See how close you come.

CHAPTER 6

A Shortcut To A Head count

The following interesting chance discovery occurred during preparations for a statistics lecture. A surprisingly good approximation to the probability of obtaining exactly n heads in $2n$ tosses of a coin is $\frac{1}{\sqrt{n\pi}}$. With this approximation available, it was then possible to obtain a more precise second approximation.

In order that we can gauge the accuracy of these approximations, we will also present the function which gives the exact probabilities. Throughout this discussion, the number of tosses will always be $2n$, so there is no need to include it in the functional notation. The notation $P(x)$ will be used to represent the probability that exactly x heads will occur in these $2n$ coin tosses. For large values of n the standard formula,

$$P(n) = \frac{C(2n,n)}{4^n}$$

proves to be quite cumbersome for hand calculator computation. Interestingly, it happens that this formula can be transformed into the relation,

$$P(n) = \frac{1 \cdot 3 \cdot 5 \cdots (2n-1)}{2 \cdot 4 \cdot 6 \cdots 2n}$$

The metamorphosis from the standard formula to the latter alternate form revealed the following number theorem.

THEOREM: For every positive integer n, 2^n times the product of the first n odd integers is equal to the product of the n consecutive integers which immediately succeed n.

On testing the accuracy of the first approximation, it was noticed that the relative error in the approximation might be varying inversely with n. Consequently, it was decided to investigate the behavior of the product n times the relative error in the approximation. Luckily, a computer analysis suggested that the preceding product did indeed approach a constant value, viz., 1/8. As a result, a second approximation to P(n) becomes the fraction

$$\frac{8n-1}{8n\sqrt{n\pi}}$$

As the table below indicates, this second approximation is exceptionally accurate. From the standard mathematical formula for coin toss probabilities, it is possible to obtain the relationships.

$$P(n-x) = P(n+x) = \frac{n(n-1)(n-2)\cdots(n-x+1)}{(n+1)(n+2)\cdots(n+x)}P(n)$$

Using these relationships in conjunction with our accurate approximation to P(n), finding the probabilities near the center of the distribution is a simple task. For example, it allows us to readily determine that, with 100 tosses of the coin, the number of heads will lie between 44 and 56 approximately 73% of the time. These central probabilities are the ones that are usually of interest.

The following table indicates the accuracy of both the first and second approximations for a few representative values of n.

n	1	10	100	250	1000
1st approx.	.564	.178412	.05641896	.035682482	.0178412412
P(n)	.500	.176197	.05634848	.035664646	.0178390111
2nd approx.	.494	.176182	.05634843	.035664641	.0178390110

CHAPTER 7

Treatise On Air Pressure

How does air pressure vary as a function of altitude? This was a question that had frequently crossed my path, but desultory searches provided unsatisfactory results. So I decided to determine the feasibility of the function's derivation.

The variables that will be used in the following discussion are: h = altitude in feet above sea level; p = air pressure in pounds per square inch at altitude h; T = the Fahrenheit air temperature at ground level; P = the air pressure at ground level; H = the altitude at ground level.

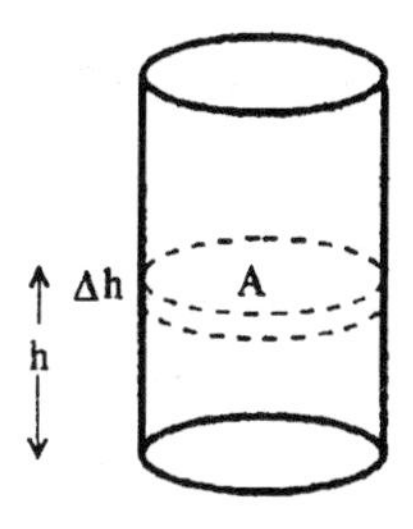

Consider a vertical column of air whose every horizontal cross section has an area of one square inch. At a point h feet above ground level we consider a slice of thickness Δh. The change in pressure, Δp, from the bottom to the top of the slice would be the negative of the weight of the air in the slice. To determine this weight, we proceed with the following steps:

1. Determine the volume of the slice of air in cubic feet. This is the area of the base, 1/144 square feet times the thickness Δh feet.

2. Multiply by the air density in pounds per cubic foot. To obtain this we refer to Avogadro's law. It implies that, at standard temperature and pressure, the weight, in pounds, of 359 cubic feet of any gas is numerically equal to the molecular weight of that gas. Since the average molecular weight of air is approximately 28.9, the weight of 359 cubic feet of air is approximately 28.9 pounds.

3. The temperature and pressure at height h will, in general, not be standard. Therefore, the density of the slice must be adjusted as described here and in step 4. The density of the air varies directly with the pressure p. Hence the air density of the slice will be the density at ground level times the factor $\frac{p}{P}$, the denominator being the air pressure at ground level.

4. The density also varies inversely as the absolute temperature. Since temperature decreases at a rate of approximately $3°F$ per 1000 feet and absolute zero is approximately $-460°F$, we must also adjust the density by the factor

$$\frac{460+T}{460+T-0.003(h-H)}$$

NOTE: For ease of reading we will temporarily substitute D for the preceding denominator: $D = 460 + T - 0.003(h - H)$

As a consequence,

$$\Delta p = -\left(\frac{\Delta h}{144}\right)\left(\frac{28.9}{359}\right)\left(\frac{p}{P}\right)\left(\frac{460+T}{D}\right)$$

$$= -\frac{0.000559(460+T)p}{PD}\Delta h$$

After separating variables and integrating, we have

$$\ln p = \frac{0.1863(460+T)}{P}\ln D + k \text{, hence}$$

$$p = k' D^{\frac{0.1863(460+T)}{P}}$$

Now we consider the situation when all measurements are at sea level and $T = 32$. In this case, $h = H = 0$ and $p = P = 14.7$. As a result, we obtain the following functional relation between p and h.

$$p = 14.7\left[\frac{460+T-0.003(h-H)}{460+T}\right]^{\frac{91.7}{P}}$$

If we let $H = 0$, that is if h measures altitude above sea level, the value of P will be 14.7 and it follows that

$$p = 14.7\left(\frac{460+T-0.003h}{460+T}\right)^{6.24}$$

Now if we ignore the temperature factor and let $H = 0$, we would obtain

$$\Delta p = -\left(\frac{1}{144}\right)\left(\frac{28.9}{359}\right)\left(\frac{p}{14.7}\right)\Delta h$$
$$= -(0.0000380p)\Delta h$$

On integrating this and using the information that $p = 14.7$ when $h = 0$, we obtain the basic formula

$$p = 14.7e^{-0.0000380h}$$

Finally this can be converted to the user friendly form:

$$p = 14.7\left(2^{\frac{-h}{18240}}\right)$$

From this form, we see that at altitude 18,240 feet, the pressure is half that at sea level. In general, at any altitude, the pressure will be halved with each 18,240 feet increase in altitude. And for each 18,240 feet down a mine shaft, the pressure will be doubled. Those who reach a depth of 10 miles will encounter air pressure of 109 pounds per square inch. For those who wonder how well the basic formula agrees with the more comprehensive one, we offer the following table.

Altitude in feet	Basic p	$T = 0°$	$T = 100°$
10,000	10.05	9.65	10.43
20,000	6.87	6.15	7.25
30,000	4.70	3.78	4.93

So for casual observation, the temperature correction may be ignored. The latter basic form of the function will do nicely.

If the comprehensive and basic forms of the pressure functions are equated for a given value of h, the result can be easily converted into a linear equation in T. Hence, for any given altitude, it is easy to determine the sea level temperature T for which the two pressure functions will agree. For example, at altitude 10,000 feet the two forms agree if $T = 48°$. The equalizing temperature at 20,000 feet would be 63°, and at 30,000 feet it would be 79°.

CHAPTER 8

Distance To The Horizon

As we sat on land and watched a ship come over the horizon, someone idly asked, "How far away do you suppose that ship is?" It was apparent that it would not be difficult to derive a formula that would apprise us of the horizon's distance. All that was needed was the solution to a simple plane geometry problem, as shown by the following diagram. Surprisingly, the solution produced a fascinating coincidence which led to a simple answer.

DIAGRAM

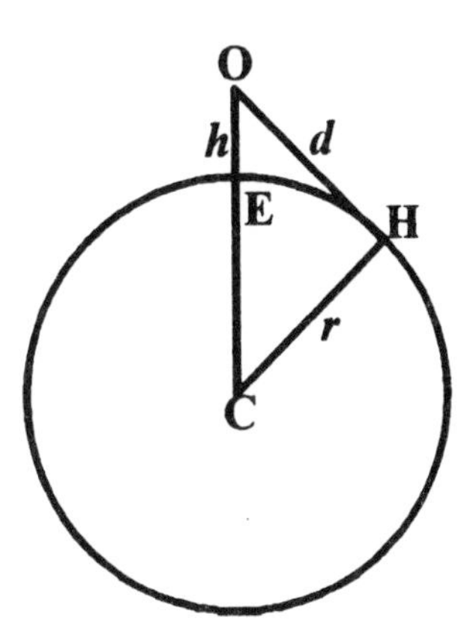

In the diagram above, the points C, O and H are respectively, the center of the earth, the position of the observer, and the position of the horizon as viewed from point O. Since the line of sight to the horizon, segment OH, is tangent to the earth's surface, angle CHO is a right angle, and thus triangle CHO is a right triangle. The distances relevant to the problem's solution are: r = radius of earth, h = length of line segment OE and represents the altitude of the observer, d = length of line segment OH and represents the distance from the observer to the horizon.

Since triangle CHO is a right triangle, the Pythagorean Theorem may be applied. It dictates that:

$$d = \sqrt{(r+h)^2 - r^2} = \sqrt{2rh + h^2}$$

Since the magnitude of h^2 pales in comparison to that of $2rh$, we may safely ignore the last term, h^2. Using 3960 miles as the radius of earth and measuring h in feet, the expression

$$\sqrt{(7920)(5280)h}$$

gives a very close approximation to the distance in feet from the observer to the horizon. Next, inside the square root symbol, we divide by 5280^2 to obtain distance, m, in miles. As a result:

$$m = \sqrt{\frac{7920}{5280} \cdot h}$$

Now for the fascinating coincidence. The fractional coefficient of h reduces to exactly 1.5. Therefore, if we measure the observer's altitude in feet, the following formula will accurately approximate m, the distance in miles to the horizon.

$$m = \sqrt{1.5h}$$

Obviously, this solution is valid only where the region between the observer and the horizon is reasonably spherical. If it is rough terrain, all bets are off.

The following two examples illustrate the accuracy of the approximation.

From an altitude of 1000 feet, the exact geometric result is 38.73030 miles and the formula's approximation is 2.4 feet less. From an altitude of 30,000 feet, the exact value is 212.208 miles while the approximation is 400 feet less.

If we compare results which this formula provides to those listed in the maritime tables, we will learn that the tabulated values are 1.08 times the corresponding values provided by the formula. The maritime tables introduction explains that the factor 1.08 stems from the refraction caused by the atmosphere.

A Morsel (food for thought)

In this era of credit, somebody must decide what the monthly payment on the loan will be. If you have access to a scientific calculator, correctly performing the numbered steps listed below will produce the value of the monthly payment.

We will be involved with the following variables:

P = the initial loan amount.

i = the monthly interest rate = the annual interest rate/12

n = the number of months during the term of the loan.

Steps:

1. After calculating i, determine the value of $1 + i$.
2. Using the scientific calculator, determine the value of $(1 + i)^{-n}$.
3. Obtain the value of $D = 1 -$ the result obtained in step 2.
4. Multiply P by i and then divide by D, (see step 3). This final result is the monthly payment.

Example: An 8% , $100,000 loan is to be paid off in 15 years.

We have i = 0.08/12 = 0.0066667, P = $100,000, n = 180. Proceeding through the listed steps should provide a monthly payment of $955.65.

CHAPTER 9

PROBABILITY ABOUT BIRTHDAYS

Once upon a time, while attending a party, I overheard two women discover that they had the same birthday. That turned out to be the basis for the investigation of the following:

PROBLEM: In a group of n people, what is the probability that at least two of them have the same birthday?

We approach the solution through the back door by first determining the probability that no two people have the same birthday. So as not to overtax the accuracy, we allow ourselves the luxury of assuming that every year contains 365 days. Those who object to this process may modify the procedure on their own time.

It is advantageous to consider the members in order. First, we notice that there are 364 chances out of 365 possibilities that the second person will not have the same birthday as the first. This makes the probability of this event 364/365. Next the probability that none of the first three has the same birthday would be the probability that the first two were unlike times the probability that the third birthday is distinct from the other two. Since there are 363 chances out of 365 for the latter situation to occur, the probability that none of the first three has the same birthday would be

$$\frac{364}{365} \times \frac{363}{365}$$

By generalizing the preceding process, we see that the probability that everyone in the group of n people has a different birthday is

$$\frac{364}{365} \times \frac{363}{365} \times \frac{362}{365} \times \cdot \cdot \cdot \times \frac{(366-n)}{365}$$

Next, we use the knowledge that one or the other of the following events is a certainty. Either at least two have the same birthday or everybody in the group has a different birthday. Fortunately we know that, if one or the other of two mutually exclusive events is a certainty, the sum of their probabilities is equal to 1. Letting P represent the probability that at least two have the same birthday, and using the preceding certainty of events, we have the equation

P + probability that every one has a different birthday = 1.

On rearranging this equation, we obtain the answer to the question.

$$P = 1 - \frac{364}{365} \times \frac{363}{365} \times \frac{362}{365} \times \cdot \cdot \cdot \times \frac{(366-n)}{365}$$

To dramatize these results, we provide the following table.

n	10	22	23	40	70	100
Probability	0.117	0.476	0.507	0.891	0.9991600	0.99999969
Odds	2 to 15	1 to 1+	1 to 1-	8.2 to 1	1189 to 1	3254689 to 1

As the table indicates, with twenty-three in the group, the odds are slightly better than fifty-fifty that there will be at least two people with the same birthday.

CHAPTER 10

A Fascinating Array Of Numbers

One day while sitting idly contemplating the intrigue of numbers, I happened to wonder if the array described below might prove interesting.

Let each term in the first row be the digit 1. Let each term in the second row be equal to the term immediately above it plus the sum of all terms preceding that term. Continue this procedure, letting every term of the array be equal to the term immediately above it plus the sum of all terms preceding that. In other words, we are considering the matrix

$[a_{ij}]$ for which $a_{1j} = 1$ and, for $r > 1$, $a_{rc} = \sum_{j=1}^{c} a_{r-1,j}$.

A section of this array is displayed below.

1	**1**	**1**	**1**	**1**	**1**	. . .
1	**2**	**3**	**4**	**5**	**6**	. . .
1	**3**	**6**	**10**	**15**	**21**	. . .
1	**4**	**10**	**20**	**35**	**56**	. . .
1	**5**	**15**	**35**	**70**	**126**	. . .
.	.	.	.	.	.	
.	.	.	.	.	.	
.	.	.	.	.	.	

Since the general term is a sum, we are obliged to discover the formula which represents that sum. The formulas for the sum of the first c terms in each of the first three rows turn out to be:

$$a_{2c} = \sum_{j=1}^{c} a_{1j} = c; \; a_{3c} = \sum_{j=1}^{c} a_{2j} = \frac{c(c+1)}{2!}; \; a_{4c} = \frac{c(c+1)(c+2)}{3!}$$

Now, we hope that the sum of the first c terms in all subsequent rows will behave similarly. Happily, mathematical induction verifies that they do. And so we have developed the following general formula:

$$a_{rc} = \sum_{j=1}^{c} a_{r-1,j} = \frac{c(c+1)(c+2)\cdots(c+r-2)}{(r-1)!}$$

Those readers for whom probability theory is not a mystery, may have noticed that the preceding function is identical to a combination function, specifically,

$$a_{rc} = C(c+r-2, r-1)$$

Using the identity $C(m,n)=C(m,m-n)$, we see that $C(r+c-2,r-1) = C(c+r-2,c-1)$. In other words, interchanging r and c does not alter the value. And since the first combination represents a_{cr}, the second combination must represent a_{rc}. This proves what our intuition might already have told us. In general, $a_{ij} = a_{ji}$. Now that we know that every number in our array is a combination number, the array displayed on the preceding page may be converted to the equivalent one shown on the next page. For the sake of consistency we define $C(0,0)$ to be equal to 1.

C(0,0)	C(1,0)	C(2,0)	C(3,0)	C(4,0)	C(5,0) . . .	C(c - 1,0)
C(1,1)	C(2,1)	C(3,1)	C(4,1)	C(5,1)	C(6,1) . . .	C(c,1)
C(2,2)	C(3,2)	C(4,2)	C(5,2)	C(6,2)	C(7,2) . . .	C(c + 1,2)
C(3,3)	C(4,3)	C(5,3)	C(6,3)	C(7,3)	C(8,3) . . .	C(c + 2,3)
C(4,4)	C(5,4)	C(6,4)	C(7,4)	C(8,4)	C(9,4) . . .	C(c + 3,4)
.	.	.	.	.	.	.
.	.	.	.	.	.	.
.	.	.	.	.	.	.

Now let us consider the sub-matrix composed of the first *r* rows and the first *c* columns. Recalling that a_{2c} equals the sum of the terms in the first row of the sub-matrix, and that $a_{r+1,c}$ equals the sum of the terms in the bottom row of this same sub-matrix, we know that the sum of the terms in column *c* from a_{2c} through $a_{r+1,c}$ would be the sum of all the terms in the rectangle. Since this would be 1 less than the sum of all the terms in column *c* down to and including $a_{r+1,c}$ and since the mentioned sum of terms from column *c* is simply $a_{r+1,c+1}$, it follows that the first term down the diagonal from the lower right corner of the rectangular array is one greater than the sum of all the terms in the rectangle. Consequently, we have the interesting identity:

$$\sum_{ij=1,1}^{r,c} C(i+j-2, i-1) = C(r+c, r) - 1$$

The insertion of combinations into the discussion steers us eventually to the binomial theorem. The path is well taken. The first row, with alternating signs, is the set of binomial coefficients for the expansion of the binomial $(a+b)^{-1}$, the second row, with alternating signs, is the set of coefficients for the expansion of $(a+b)^{-2}$. Yes, it can be shown that row *r*, with alternating signs, is the set of coefficients for the expansion of $(a+b)^{-r}$.

Next, we turn our attention to the sum of the terms along the diagonal from a_{r1} through a_{1r}. By happenstance, we have hit a jackpot. The sum is equal to $2^{(r-1)}$, and this provides a recollection and a surprise. Our array is simply the rotated Pascal's Triangle.*

This provides information that I, at least, did not previously have. Pascal's Triangle supplies sets of binomial coefficients for all integral exponents, not just the non-negative exponents.

Finally, every square matrix which borders on the boundaries of the array, that is, for which $a_{i1} = 1$ and/or $a_{1j} = 1$, has a determinant equal to 1.

I suspect that the array has other interesting characteristics. Help yourself, then let me know.

* Pascal's Triangle: 1's on the outside; each remaining number is the sum of the two closest numbers above it. The 2nd row of this triangular array displays the resulting pair of coefficients generated when the binomial (a+b) is raised to the 1st power. The 3rd row displays the sequence of coefficients generated when the same binomial is raised to the 2nd power. The 4th row displays the sequence of coefficients generated when the same binomial is raised to the 3rd power, etc.

1
1 1
1 2 1
1 3 3 1
1 4 6 4 1
1 5 10 10 5 1

CHAPTER 11

*What Goes Up Must Come Down

Interest in the following topic was triggered by a rash of small arms attacks that had recently been directed at overhead aircraft. Someone had asked, "How high will a rifle bullet go?" The question was the motivation for this investigation to determine the altitude that can be reached by a bullet fired vertically from the ground.

The principle parameters considered to be relevant were muzzle speed, bullet weight, ground altitude and ground temperature. Fortunately, data found in a popular gun magazine kept the problem from remaining in the realm of the purely academic. A table displaying bullet speeds at various sea level distances from the muzzle made it possible to discover the air resistance encountered by a bullet in flight. It was known that, at high speeds, the air resistance is proportional to the square of the speed. The tabular values produced the proportionality constant required, and it was the key to the solution.

Data for a 150 grain bullet were chosen and solutions reflecting two different ground altitudes and two different ground temperatures were obtained with computer assistance. But a solution without verification is not completely satisfying – it requires confirmation.

Of course it was impossible to ascertain the altitude attained by an experimental bullet, but the time of flight could be observed. If the observed times were in substantial agreement with predicted values, then it would be assumed that the experimental altitude should also be in agreement.

In order to test the theoretical results, it was decided to use a 30-06 rifle with a peep sight. A plumb line was securely taped to the knife-edge front sight, passed through the peep opening and attached to a plumb bob. After taping the muzzle of the rifle to a stake, the butt of the rifle was adjusted until the plumb line was free in the rear-sight opening, at which time the rifle was discharged. Contact with the weather bureau had assured us that wind speeds were negligible. Normally the line of sight, i.e., the direction of the plumb line, would not be exactly parallel to the bore of the rifle, but it was decided that it was acceptably close. Observers, heads and shoulders well shielded, timed three separate shots and were rewarded by three thumps as each bullet returned.

This experiment, performed at sites of distinctly different altitudes in Colorado and Pennsylvania, provided times that were six to seven per cent greater than those that had been predicted. One reason for this was suggested by the warbling sound the bullets made as they returned to earth. This indicated that the bullets had stopped spinning and were tumbling on the final part of their flights. This should have been anticipated, but wasn't. Then too, there was no knowledge as to the validity of the muzzle speed for the ammunition used. It had been supplied by the gun manual.

Page 45 contains the table which presents four examples of the solution. Two different ground altitudes and two different ground temperatures are considered. In each case bullet speeds are shown, for both the rising and the falling bullet, at 300 foot intervals. Notice that the solution indicates that the falling bullet reaches a maximum speed when its altitude is in the neighborhood of 2000 feet, then gradually decreases before it strikes the ground. By this time, the density of the air is sufficient to provide air resistance slightly greater than the weight of the 150 grain bullet.

It happens that the table supplies information which enables us to quickly determine the force with which the air resists the bullet as it leaves the muzzle. As our table indicates, during ground approach the bullet speed is nearly constant. Hence the air resistance must be approximately equal to 1 gravity. As previously mentioned, physicists know that air resistance at these speeds varies as the square of the speed. So a first approximation to the air resistance at the muzzle, measured in gravities of force, would be simply the square of the ratio between the muzzle speed of the bullet and the speed at which the bullet strikes the ground. Two of these muzzle air resistance values, the ones at opposite ends of the temperature-altitude spectrum, are 68.5 gravities and 95.8 gravities. The preceding results are based on a presumed zero acceleration at ground approach, but the table indicates that this acceleration is slightly negative. If we use the tabulated downward bullet speeds at altitudes 300 feet and 0 feet, it is possible to refine this estimate of the small deceleration and the resultant air resistance at ground approach. From these tabulated downward speeds we can determine an estimate of the average speed during the final 300 feet descent and from this we can determine the time required for the descent. Finally, the ratio of change in speed to time of descent produces a second approximation to the ground approach deceleration and air resistance. We will use the bullet at altitude 1140 feet and temperature 0 degrees as an example. The respective downward speeds are 304.7 and 303.5 ft/sec. Hence the average speed is 104.1 ft/sec. and the elapsed time for the 300 foot descent would be 0.987 seconds. If the 1.2 ft/sec change in speed is divided by this time interval, the deceleration is slightly more than 1.2 ft/sec^2. Dividing this result by the acceleration of gravity indicates that the deceleration is approximately 0.038 gravities instead of 0 gravities, and hence the ground approach air resistance must be 1.038 gravities instead of 1 gravity. As a result, the second approximation for the muzzle force in gravities is 1.038 times the first approximation.

For example, that muzzle air resistance which was previously estimated at 95.8 gravities becomes nearly 100 gravities. So the force against the 150 grain bullet is initially nearly 100 times 150 grains, or slightly more than 2 pounds. Hence there is ample reason for the large difference between the heights attained by a bullet in atmosphere and in a vacuum. In atmosphere, the low altitude-low temperature bullet reaches a calculated height of 6851 feet. Any object projected vertically upward in a vacuum will reach a height equal to the square of its speed divided by twice the acceleration due to gravity. Since our experimental bullet had a muzzle velocity of 2970 ft/sec, it would reach a height of approximately 26 miles in a vacuum.

GROUND ALTITUDE 6110 feet					**GROUND ALTITUDE 1140 FEET**			
Ground Temp. 90 F		Ground Temp. 0 F			Ground Temp. 90 F		Ground Temp. 0 F	
Speed Up	Speed Down	Speed Up	Speed Down	Altitude	Speed Up	Speed Down	Speed Up	Speed Down
Maximum Elev. 8877'		Maximum Elev. 8092'			Maximum Elev. 7772'		Maximum Elev. 6851'	
108.8	104.9			8700				
184.4	166.9			8400				
243.2	206.6			8100				
296.3	236.1	141.9	132.7	7800				
347.8	259.2	209.8	182.6	7500	135.9	127.9		
397	277.8	266.9	216.8	7200	206.6	178.9		
447.8	293.1	320.7	242.5	6900	265.6	213.1		
498.6	305.8	373.4	262.6	6600	321.1	238.5	131.9	122.6
551.6	316.3	426.5	278.7	6300	f375.8	258.1	204.6	174.1
606.7	325.2	480.8	291.8	6000	431.1	273.7	266.8	207.6
664.3	332.6	537.1	302.5	5700	488.2	286.3	326	231.7
724.9	338.8	596.2	311.2	5400	547.7	296.4	385.4	250
788.9	343.9	658.5	318.3	5100	610.4	304.6	446.7	264
856.8	348.2	724.8	324.1	4800	677	311.3	511	275
929.1	351.7	795.2	328.8	4500	748.1	316.7	579.4	283.6
1006	354.6	870.9	332.6	4200	824.3	321	652.9	290.3
1088.2	357	952.2	335.6	3900	906.4	324.5	732.5	295.5
1176.4	358.8	1039.9	337.9	3600	995.1	327.2	414.1	299.5
1270.9	360.2	1134.8	339.6	3300	1091.3	329.3	913.9	302.5
1372.5	361.2	1237.7	340.8	3000	1195.7	330.8	1018.1	304.7
1481.9	361.9	1349.5	341.6	2700	1309.3	331.9	1133.8	306.2
1599.9	362.3	1471.3	342	2400	1433.3	332.6	1260	307.2
1727.2	362.5	1604.1	342	2100	1568.8	333	1400.9	307.7
1864.8	362.4	1749.3	341.9	1800	1717.2	333.1	1557.6	307.8
2013.7	362.2	1908.1	341.4	1500	1879.7	333	1732.9	307.6
2175.3	361.8	2082.3	340.8	1200	2058.2	332.7	1926.8	307.2
2349.9	361.2	2273.5	339.9	900	2254.3	332.2	2144.6	306.5
2539.7	360.6	2483.7	339	600	2470.2	331.5	2388.7	305.7
2746	359.7	2715.2	337.8	300	2708	330.7	2662.5	304.7
2970.3	358.8	2970.3	336.6	0	2970.3	329.8	2970.3	303.5
Time Up	Time Down	Time Up	Time Down		Time Up	Time Down	Time Up	Time Down
16.9	31.7	16	30.4		15.6	30	14.5	28.4

A Morsel (food for thought)

Suppose an object is propelled in a vertically upward direction from the surface of one of the universe's atmosphereless globes. If its initial speed is v, elementary physics teaches us that the maximum height attained by the object will be $\frac{v^2}{2g}$, where g represents the acceleration of gravity at the globe's surface. But this result assumes that the acceleration of gravity remains constant during the object's flight. In actuality, g varies inversely as the square of the object's distance from the center of the globe involved. This implies a decrease in the average value of g, and hence the object reaches a height greater than that predicted by the elementary formula. Hopefully you are now wondering how much these considerations increase the maximum height attained by the rising object. The answer is below. In the following expressions, h is the maximum height under constant gravity and R represents the radius of the relevant globe:

h *is* $\frac{v^2}{2g}$ and the increase in height due to decreasing g *is*

$\frac{h^2}{(R-h)}$. The next three comments put this result into perspective. For these examples, the relevant globe is the earth.

1. If the initial vertical speed is 100 ft/sec, the increase to the basic 155 ft height is 0.35 millimeters.

2. If the initial vertical speed is 3000 ft/sec, the increase to the basic 26.5 mile height is 943 ft.

3. If the initial vertical speed, v, happens to be equal to or greater than $\sqrt{2gR}$, there's no way you can imagine how large the increase is. The object is gone for good.

CHAPTER 12

*RAMIFICATIONS OF LEAP YEAR

Maybe you have wondered on what day of the week you were born, what day of the week a specific date occurred in the past, or what day of the week a specific date will occur in the future.

As we advance from one year to the next, the day of an anniversary date advances through the week for the following two reasons. Our regular year of 365 days contains 52 weeks plus 1 day. As a result, an anniversary date advances one day through the week for each passing year. If, in addition, there is an intervening leap day, Feb. 29, an anniversary date will advance an extra day. Therefore, over a span of years, the total number of days of advancement would be the sum of the number of years elapsed plus the number of intervening leap days. If this total happens to be a multiple of 7, the anniversary date will have advanced exactly that multiple of weeks. In this case the initial day of the week will be identical to the final anniversary day of the week. This implies that the number of days advanced over a number of years is the same as the remainder obtained when the aforementioned total is divided by 7. The value of the remainder, when the total days of advance is divided by 7, tells us that the anniversary date has proceeded through a certain number of weeks plus the number of days indicated by the remainder.

In other words, to determine the day of the week on which an anniversary date falls, follow these steps:

1. Determine the day on which it falls in the current year.

2. Determine the number of years between the current year and the one in question.

3. Determine the number of intervening leap days and total it with the number obtained in the preceding step.

4. Find the remainder when the preceding total is divided by 7.

5. If the year in question is in the future, advance the current anniversary day by this remainder number of days. If, instead, the year is in the past, subtract this remainder number of days.

As a consequence, it can be shown that everybody's 28th birthday falls on the same day of the week as did the day of birth. If the birthday is not Feb. 29, the 28th birthday will be the fourth time that the day of the birthday and the day of birth have been on the same day of the week. This verifies what you may have already suspected about anniversaries that do not fall on Feb. 29. If an initial anniversary date falls on a certain week day, then over the long run all subsequent anniversaries will tend to fall on this same week day one time in seven. But if the anniversary date is Feb. 29, the 28th anniversary is the first one for which the week days coincide.

The cycle of intervening years in the aforementioned 28 year cycle follows the order 5, 6, 11, 6. The starting number in this cycle depends on the portion of the leap year cycle in which the date of birth fell. As an example, those born in a leap year, but after Feb. 29, start at the first 6 in the cycle and hence the order for them is 6, 11, 6, 5, which means that birthdays 6, 17, 23, 28, etc., fall on the same day of the week as was the day of birth.

If the span of years includes a century year, the following information is necessary. If the number of a year is evenly divisible by 4, the year is a leap year, unless it is also divisible by 100, in which case it is not a leap year, unless it is also divisible by 400, in which case it is a leap year, unless it is also divisible by 4000, in which case it is not a leap year. For example, the years 1900, 2100, 2200 and 2300 are not leap years, but 2000 and 2400 are leap years.

The following example applies information just offered.

Given that the date of Lincoln's Gettysburg Address was Nov. 19, what day of the week was it? Lincoln, referring to the Declaration of Independence, said, "Four score and seven years ago". So we know that the address was given on Nov. 19, 1863. First, we must determine the anniversary week day in the current year. Assume that we were solving this problem in the year 1997. That year Nov. 19 occurred on Wednesday. Since 1900 was not a leap year, there had been only 33 leap days since 1863. When we add 33 to the number of intervening years, 134, we obtain a total of 167. Then, division of 167 by 7 produces a remainder of 6. Hence we go back 6 days, or forward 1 day, and we learn that the Gettysburg Address was on Thursday.

If leap year occurred every 4 years, we should infer that the length of the year is 365.25 days. Of course, we know that the year's length is not precisely that, but how close is it? A very accurate answer is obtained by resorting to the rule defining leap year. Accordingly, there are 30 century years, in addition to the 4000th year, which are not leap years. Consequently, during a 4000 year cycle there are 1000 – 31 = 969 extra days, or 0.24225 extra days per year. On this basis, the year contains 365 days, 5 hours, 48 minutes, 50.4 seconds. This is 4.4 seconds longer than the length of year listed in the almanac.

Since the length of our year is slowly increasing, this discrepancy will decrease for a few years. Ultimately, the error will then increase and we will have to adjust the rule for determining which years are leap years. So now we have something else to worry about.

A Morsel (food for thought)

Have fun with this parlor trick. Consider any number containing more than one digit. If the sum of the digits of that number is subtracted from that number, the sum of the digits in the resulting difference will always be a multiple of 9.

Ask the participant to perform the aforementioned subtraction without informing you of the result. Then ask the participant to circle one of the digits in the difference and then tell you the sum of the remaining digits of that difference. Since you know that the sum of all the digits is a multiple of 9, you will be able to tell the participant the value of the digit circled. If they disagree with your solution, ask them to recheck their work. For example, if the remaining sum is 7 you will know that they must have circled the digit 2. If the remaining sum is 13, the digit 5 must have been circled. If the remaining sum is 26, the digit 1 must have been circled. Remember that in each case the sum of all digits must be a multiple of 9. It may happen that the remaining sum is itself a multiple of 9. In this case there are two possibilities. The digit circled may be either 0 or 9.

CHAPTER 13

*MISCELLANEOUS TIDBITS

TREATISE ON A CC OF $H_2 0$:

Wonderment about the unfathomable minuteness of a molecule prompted the arithmetical research that provided the following fascinating trivia.

To better conceptualize the minuscule molecule, it was decided to determine what volume a cubic centimeter of water would enlarge to if each of its molecules was transformed into a grain of wheat.

According to a law of Avogadro, the number of molecules in a gram-molecular weight of an element or compound is 6.02×10^{23} molecules. The molecular weight of water is 18, and as a consequence, one cc of water, which weighs one gram, contains 3.34×10^{22} molecules. Next, since we know that one grain, in our system of weights, originated with the weight of a grain of wheat, one pound of wheat would contain approximately 7000 grains. Since one bushel of wheat weighs 60 pounds and one bushel contains almost 1.25 cubic feet, one cubic foot of wheat contains approximately 337,000 grains of wheat. With these figures, the volume of the enlarged molecules would be 9.92×10^{16} cubic feet. Armed with this information, we divide this result by the area of the 48 contiguous states, 2,957,000 square miles, and learn that the enlarged molecules would cover the 48 states to a depth of 1200 feet. As staggering as the latter information is, the following calculations may be a topper.

Suppose we line up all the molecules in a cc of water along a straight line. How far, in miles, will they extend? To answer this question, we must first determine the diameter of a water molecule. If we temporarily, but incorrectly, assume that the molecules are spheres stacked one atop another in identical layers, the number of molecules that would be lined up along the 1 centimeter edge would be the cube root of the total number of molecules. This turns out to be 3.22×10^7, and therefore the diameter of the molecule would be 3.11×10^{-8} centimeters, or 3.11 angstroms. (1 angstrom = 10^{-10} meters). In actuality the molecules would not be stacked one atop another, but nested between the molecules in the layer below them, as are cannon balls when stacked. This stacking adjustment would decrease the volume from the original 1 cc to 0.866 cc. Thus, in order to expand the volume to the original 1 cc, the diameters would have to be increased to 3.26 angstroms.

So this latter measurement would be the diameter of a water molecule if the molecules were spheres packed tightly in a 1 cc cube. Not surprisingly, the physicists inform us that this result is incorrect. They say that the water molecule's diameter is 4.6 angstroms. So why the discrepancy? After much investigation the answer is apparent. The discrepancy is caused by an anomaly.

Now we wend our way as follows: Multiply the total number of molecules by the diameter of each molecule to obtain the total distance of 1.54×10^{15} centimeters. Then divide this by 2.54 cm/in, by 12 in/ft, by 5280 ft/mile to get the final result of 9.56×10^9 miles. Finally we divide by the distance from the earth to the sun, 93,000,000 miles, and we are forced to accept that which verges on the unbelievable. If the molecules from a cubic centimeter of water were lined up along a straight line, the infinitesimal molecules would extend approximately 103 times the distance to the sun. If you prefer, this is equivalent to a distance of 14.2 light hours.

A PRIME PROPERTY:

Consider the five integers between any two successive multiples of 6. Since any multiple of 6 must have both 2 and 3 as factors, we know that the second and fourth of the five integers must have factors of 2 while the third of the five integers must have a factor of 3. Consequently, the only integers of the five that can be prime are the first and the fifth. Thus we have the following theorem: In order that an integer be prime, it is necessary that it be adjacent to a multiple of 6.

Of course this condition is not sufficient, but, at least initially, many of those integers adjacent to a multiple of 6 are prime. In fact, each of the first 19 multiples of 6 have at least one prime neighbor, and in 9 of those cases, both neighbors are prime. There are many occurrences of so-called prime twins, that is, pairs of primes that are separated by a composite integer. In every instance, the intervening composite will be a multiple of 6.

THE CIRCLE SHEDS ITS SECRET:

One of the concepts which repeatedly mystified me during my pre-calculus days, was that of the creation of the number π. I realized that the accuracy of the number then representing it was far beyond the capability of direct measurement, so how could we calculate the precise value? The exhilaration I experienced when calculus taught how it could be done will never be forgotten. The fact that human brains had been equipped to, with only an idealized circle, conquer this abstraction π and make its evaluation procedure comprehensible was a spiritual experience.

All who have a good background in calculus have learned the following simple infinite series representation for $\frac{\pi}{4}$.

$$\frac{\pi}{4} = 1 - \frac{1}{3} + \frac{1}{5} - \frac{1}{7} + \frac{1}{9} - \frac{1}{11} + \cdots + \frac{(-1)^{(n+1)}}{2n-1} + \cdots$$

The sum of an even number of terms is always less than $\pi/4$ and the sum of an odd number of terms is always greater than $\pi/4$. Two successive sums always bracket the value of $\pi/4$. The lengths of the brackets in this sequence approach zero as the number of terms in the sum increases. This particular series is inefficient, since it requires the addition of an oppressive number of terms in order to get sufficient accuracy. But the only way to find the value of π is by means of an infinite series.

AN ATOMIC CONJECTURE:

For elements that have been discovered as of this date, the numbers of electrons required to complete the successive shells are 2, 8, 8, 18, 18 and 32. Since the numbers of the sequence 2, 8, 18 and 32, can be expressed as 2×1^2, 2×2^2, 2×3^2, 2×4^2, the author conjectures that the first future shell which contains more than 32 electrons, will contain 50 electrons. Most of those to whom the author has expressed this conjecture, have stated that they thought the sequence phenomenon to be a coincidence. The author's gut feeling is, "No way!" Who will prove the conjecture right or wrong?

CHAPTER 14

*The Cinematic Stroboscope

If you have watched old-time Western movies and wondered about the bizarre behavior of the rotating wagon wheels, you will find the explanation here. In case you have forgotten precisely what happened, the following will refresh your memory.

Assume that we are watching a motion picture of the following scene. A horse drawn carriage is traveling to the right with wheels rotating at 2.4 revolutions per second (rps). Its wheels contain 12 evenly spaced spokes, 30^0 apart. Then the buggy gradually slows and eventually stops. During this process, the wheel will proceed through the following five phases.

1. Initially, the wheel which is rotating clockwise at the actual rate of 2.4 rps, will appear to be rotating clockwise, but at the rate of 0.4 rps.

2. As the buggy wheel slows to 2 rps, the projected wheel will gradually slow and temporarily appear to be stationary.

3. As the actual wheel slows and approaches a speed of 1 rps, the motion picture wheel will rotate counterclockwise (backward), and its speed will apparently increase and approach 1 rps.

4. When the actual speed of rotation arrives at 1 rps, the projected wheel will display a series of sudden convulsive jumps with no sense of direction.

5. As the buggy then comes to a stop, the projected image will reflect the true speed and direction of the wheel's motion.

The discourse which follows is presented for those readers who are interested in knowing why all this happens. To make the phenomenon understandable, we consider the following situation. Suppose we have a wheel containing 12 spokes spaced at 30° intervals, attached to a mechanism which allows the wheel to be rotated at any desired speed. This mechanism is in a room whose only source of light is a device which emits very short flashes of light at time intervals which are adjustable to any desired length. Such a device is called a stroboscope. For reasons that will become apparent later, we initially choose a wheel speed of 2.4 rps, in a clockwise direction, and 1/24 second intervals between flashes. Consequently, the strobe will initially flash 24 times per second and the wheel will rotate 36° between flashes. In the following diagram the letters, which label the spokes, enable us to follow the progressive changes in spoke positions at successive flashes. The three positions in the diagram will be used as reference for the discussion which immediately follows.

DIAGRAM 1

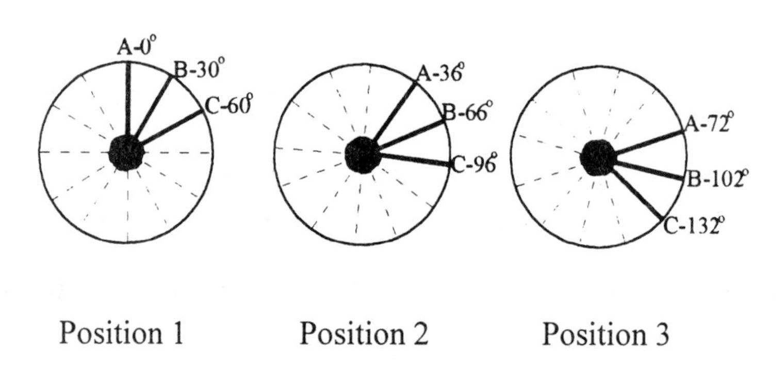

Acceptance of the following principle is critical to the understanding of the conclusions that will be developed. As the eye follows the changing position of the wheel from scene to scene, the brain will assume that a given spoke moves from its first position to the location of that spoke which arrives nearest to that position in the next scene. With spokes 30° apart, the nearer spoke will be that one which arrives at a point less than 15° distant. Recalling that the wheel is rotating clockwise at 2.4 rps, equivalent to 36° per flash, we will assume that, at the first flash, the wheel is at position 1 as diagrammed on the preceding page, then at position 2 for the second flash, and at position 3 for the third flash.

Notice that, at the time of the first flash, spoke C appears at the 60° position. Then at the time of the second flash, spoke B appears at the 66° position. Finally, at the time of the third flash, spoke A appears at the 72° position. During that three flash interval, no other spoke will be seen that close to the initial position of spoke C. So, during that 1/12 second interval, the brain will discern that a single spoke has moved from the 60° position to the 72° position. The brain will then interpret that spoke C, and hence the entire wheel is rotating clockwise at the rate of 144 degrees per second or at a rate of 0.4 rps. Remember that the wheel is actually rotating clockwise at a rate of 2.4 rps.

Now suppose that this rotating wheel had been photographed by a standard old-time movie camera. Since that movie camera exposed frames at the same familiar rate of 24 per second it will record, and later project, the same scenes that our eyes would perceive if those scenes were defined by the stroboscope.

Attention will now be focused on what transpires at the different wheel speeds of 2 rps, 1.8 rps, 1 rps and 0.8 rps.

First consider the wheel to be rotating at 2 rps, equivalent to 30° per flash, and with the spokes at the numbered clock positions, which are also 30° apart. With each successive flash, each spoke will have advanced to its next clock number. As long as the wheel is rotating at 2 rps, every frame of the film indicates the same apparent position for the wheel. Hence, the movie version shows a stationary wheel. Next, consider the situation when the wheel has slowed to 1.8 rps. Now each spoke will advance 27° between frames. Thus, at the initial flash spoke C will be at 60°, at the second flash spoke B will be at 57° and at the third flash, spoke A will be at 54°. This is shown in the following diagram.

DIAGRAM 2

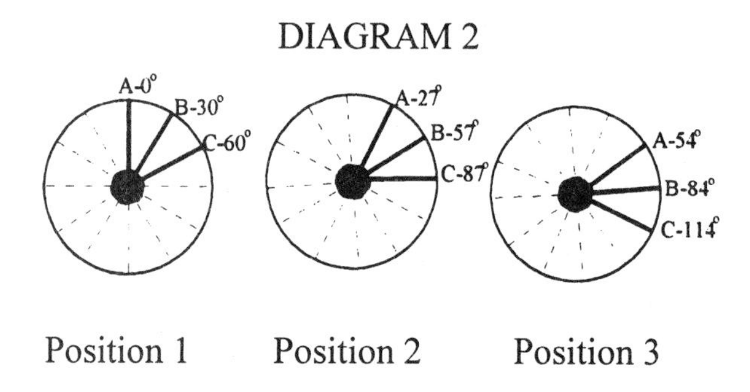

Position 1 Position 2 Position 3

Again notice that at the time of the first flash, spoke C appears at the 60° position. At the time of the second flash, spoke B appears at the 57° position. Finally at the time of the third flash, spoke A appears at the 54° position. During the time of the three flashes, no other spoke will be seen that close to the initial position of spoke C. So during that 1/12 second interval, the brain will discern that a single spoke has moved from the 60° position to the 54° position. The brain will then interpret that spoke C, and hence the entire wheel, is rotating counterclockwise (backward) at the rate of 72° per second or 0.2 rps.

When the wheel speed reaches exactly 1 rps, or 15° per flash, the spokes will alternate between the positions shown in Diagram 3.

DIAGRAM 3

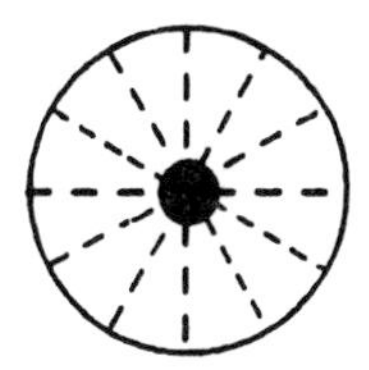

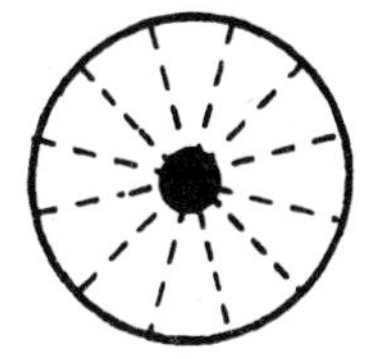

Position 1 Position 2

Each flash will reveal that the position of each spoke has suddenly jumped from its preceding position to a position halfway toward one or the other of its neighboring spokes. The brain cannot discern any sense of direction from these scenes and sees only a series of convulsive random jumps one way or the other.

Finally, it can be shown that when the wheel is moving slower than 1 rps, the eye will see the true positions of the spokes and the perceived motion will be the true motion. This perception continues until the wheel comes to its actual stop.

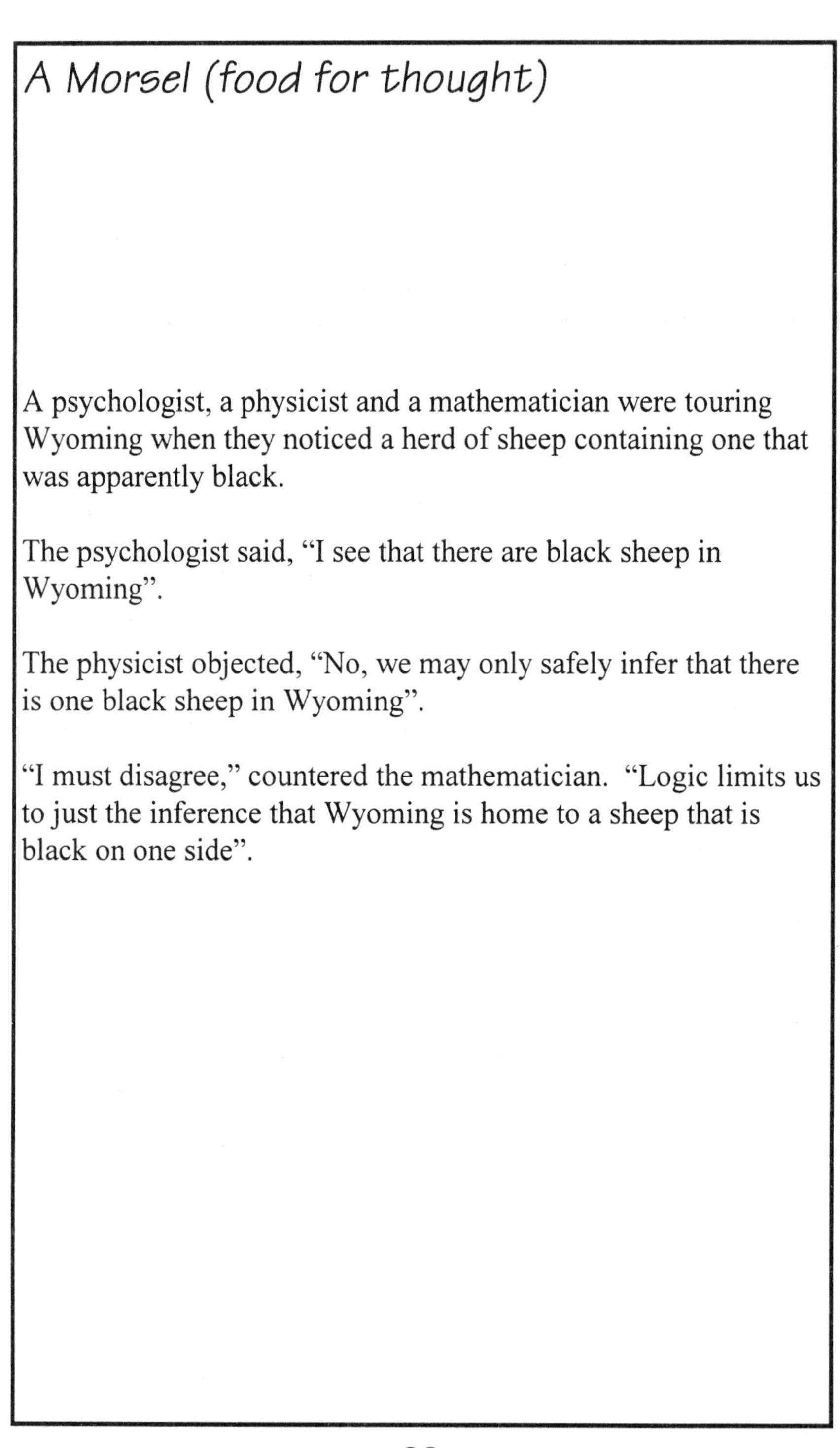

A Morsel (food for thought)

A psychologist, a physicist and a mathematician were touring Wyoming when they noticed a herd of sheep containing one that was apparently black.

The psychologist said, "I see that there are black sheep in Wyoming".

The physicist objected, "No, we may only safely infer that there is one black sheep in Wyoming".

"I must disagree," countered the mathematician. "Logic limits us to just the inference that Wyoming is home to a sheep that is black on one side".

CHAPTER 15

*Tips For Mental Arithmetic

Follow the tips listed below and you may surprise, maybe even amaze, your friends unless they are already accustomed to your brilliance. Of course, the first requisite is a facility for arithmetic. Given that, mastery of the following principles combined with plenty of practice can work wonders.

The simplest hint is involved in the following types of multiplication examples. Multiply 102 by 33. Since this may be interpreted as being the sum of 102 33s, we simply add two 33s to one hundred 33s. So the result must be 66 plus 3300, or 3366. The same reasoning would apply if we were required to determine the value of 19 times 24. Nineteen 24s would be one 24 less than twenty 24s. In other words the result is 480 minus 24, or 456.

The squares of integers appear so frequently when performing mental arithmetic, that the squares of integers from 1 through 25 should be committed to memory.* They are particularly useful in the application of the following rule of elementary algebra:

$$(a + b)(a - b) = a^2 - b^2.$$

This rule applies whenever two numbers that must be multiplied are equidistant from a third number. For example:

$$19 \times 21 = (20 - 1)(20 + 1) = 20^2 - 1^2 = 399$$

$$45 \times 35 = (40 + 5)(40 - 5) = 40^2 - 5^2 = 1600 - 25 = 1575$$

*Interestingly, the squares of integers between 25 and 50 have the following trait:

$(25 + n)^2 = 100n + (25 - n)^2$. Similarly, $(50 + n)^2 = 200n + (50 - n)^2$, and $(75 + n)^2 = 300n + (75 - n)^2$, etc.

Examples: $24^2 = 576$ and $26^2 = 676$; $23^2 = 529$ and $27^2 = 729$;
$49^2 = 2401$ and $51^2 = 2601$; $48^2 = 2304$ and $52^2 = 2704$.

Determine the number of seconds in a day. In one hour there are
$60 \times 60 = 3600$ seconds, so in one day there would be
$24 \times 3600 = 100 \times (30 - 6)(30 + 6) = 100(900 - 36)$
$= 86{,}400$ seconds

The rule also works in reverse. If the difference $38^2 - 37^2$ is sought, we convert to $(38 + 37)(38 - 37) = 75 \times 1 = 75$. This example informs us that the difference in the squares of two consecutive integers is simply the sum of those integers.

Another useful rule is $(a \pm b)^2 = a^2 \pm 2ab + b^2$. For example:
$29^2 = (30 - 1)^2 = 900 - (2 \times 30 \times 1) + 1 = 841$

$52^2 = (50 + 2)^2 = 2500 + (2 \times 50 \times 2) + 4 = 2704$

$1012^2 = (1000 + 12)^2 = 1{,}000{,}000 + 24{,}000 + 144 = 1{,}024{,}144$

If a number ends with 5, its square is especially simple. Delete the 5, multiply the remaining number by 1 greater than that number, and then annex 25.** For example:

$65^2 = 6 \times 7$ and annex $25 = 4225$

$125^2 = 12 \times 13$ and annex $25 = 15{,}625$

Sometimes more than one procedure will be required. For example,

$$92 \times 98 = (95 - 3)(95 + 3) = 95^2 - 3^2 = 9025 - 9 = 9016$$

**For those who are interested, a verification for the following procedure is given at the end of this chapter.

In the beginning, when each step has to be contemplated, the entire process will be no short-cut, but practice eliminates confusion and brings speed.

Algebraic techniques are also helpful with problems to which the preceding methods do not apply.

Example: 61 times 82 may be converted to (60 + 1)(80 + 2).

To solve this, we add the four products obtained when each term in the first parentheses is multiplied by each term in the second parentheses, which gives 4800 + 80 + 120 + 2 = 5002.

Perhaps one of the most valuable tools is a knowledge of aliquot parts, that is, the decimal equivalents of the simpler common fractions as partially displayed in the following table.

1/2 = .50			
1/3 = .33 1/3			
1/4 = .25	3/4 = .75		
1/5 = .2			
1/6 = .16 2/3		5/6 = .83 1/3	
1/8 = .125	3/8 = .375	5/8 = .625	7/8 = .875
1/9 = .11 1/9			
1/11 = .09 1/11			
1/12 = .08 1/3			
1/15 = .06 2/3			
1/16 = .0625	3/16 = .1875	5/16 = .3125	7/16 = .4375

Many of the multiples of the entries in the first column are not listed because they are easily determined as needed. As we know, some of these aliquot parts have a final digit, such as 1/8 = 0.125. Others such as 1/3 = 0.333 . . . , do not have a final digit, but repeat a certain cycle of one or more digits ad infinitum. . . Such examples are called repeating decimals. The aliquot part 1/7 = 0.142857142857 . . . was not included, but it is particularly interesting. Amazingly, every proper fraction with denominator 7, has a decimal equivalent with this identical order of repeating digits, but starting with a different digit in each case. If we start with the 2, that is, 0.2857142857 . . . , we have 2/7, if we start with the 4, and keep the same order, we have 3/7, if we start with the 5, we have 4/7, if we start with the 7, we have 5/7, if we start with the 8, we have 6/7.

When working with fractions, it is wise to routinely reduce them to lowest terms. For example, suppose we are interested in obtaining the batting average of a player who has received 27 hits in 63 times at bat. The batting average would be the decimal equivalent of the fraction 27/63. Reduced to lowest terms, the fraction transforms into 3/7. If we have memorized the order of the digits for fractions with denominator 7, we would know that 3/7 = 0.429, after the final digit has been rounded off.

Mental arithmetic often requires complete disregard for decimal points until all operations have been performed. Attention to them simply muddies the waters. Disregard of the decimal points does not alter the digits in the answer and, if the problem is not simple, it makes the digits much easier to determine. Common sense will usually quickly determine the obvious position for the answer's decimal point. Also, judicious multiplication by extraneous powers of 10 is often a handy catalyst. The preceding two tips are involved in some of the examples which follow.

Given that one pound avoirdupois is equivalent to a weight of 7000 grains, how many grains are there in one ounce? Division of 7000 by 16 requires the decimal equivalent for 7/16 and common sense tells us that the answer should be slightly less than half of 1000. So the answer must be 437.5. Incidentally, one pound troy weight contains 5760 grains. Since there are only 12 ounces in a troy pound, one troy ounce contains 480 grains. Hence, an ounce troy is heavier than an ounce avoirdupois, but a pound troy is lighter than a pound avoirdupois.

The question, "Which is heavier, a pound of feathers or a pound of gold?" is not a nonsense question. We see that the correct answer is a pound of feathers, since feathers are measured in avoirdupois and gold is measured in troy weight. The pound of feathers weighs 7000 grains, but a pound of gold weighs only 5760 grains.

An acre is what decimal fraction of a square mile? Since there are 640 acres in a square mile, we need the decimal equivalent of 1/640. Since we don't know this one, we use the fact that the digits in the decimal equivalent 1/640 has the same digits as 100/64. We use the latter fraction to determine the digits and then decide where the decimal point goes. The fraction 100/64 reduces to 25/16, which is 1 and 9/16. Recalling the aliquot part for 9/16, gives the correct digits, 15625, and common sense indicates a result somewhat larger than 1/1000. Therefore, 0.0015625 is our answer.

Given that water weighs 62.4 pounds per cubic foot, what is the weight of water in a 48 cubic foot tank? After we are acquainted with our aliquot parts, we will recognize that the digits in 62.4 differ only slightly from those of the digits representing the decimal for 5/8, viz., 625. So we multiply 5/8 by 48 cubic feet, let common sense position the decimal point, and obtain the result, 3000 pounds. But this assumes 62.5 instead of 62.4 pounds per cubic foot. Therefore we must subtract 0.1 pound times 48 and obtain the final result, 2995.2 pounds.

If the cube of 15 is required, we cube 3/2 instead, since $15 = (3/2) \times 10$. $(3/2)^3 = 27/8 = 3 + 3/8$. With the aliquot part for 3/8, and proper positioning of the decimal point, we find that 15 cubed is 3375. Similarly, 25^3 converts to determining $(5/2)^3$ or 125/8 or 15 + 5/8 which leads to a result of 15,625.

Since 3/4 and 75 are interchangeable, for our purposes, the problem 44×75 reduces to $44 \times 3/4$, which leads to the answer 3300. 25×37 leads to $1/4 \times 37$. Next, $1/4 \times 37 = 37/4 = 9 + 1/4$. Thus $25 \times 37 = 925$. 212/25 leads to 212 divided by 1/4 or 4 times 212, which then leads us to the answer 8.48.

Likewise, 741/50 becomes 741 times 2, and this suggests the answer 14.82. Finally, $(1/15)^2$ becomes, after 15 is replaced by 3/2, 4/9 and this ultimately implies the result 0.00444 . . . .

Some of the examples shown have suggested these helpful hints. If multiplication (or division) by 5 is required, you might prefer instead to divide (or multiply) by 2. Also, if it is required to multiply (or divide) by 25, it is simpler to divide (or multiply) by 4.

The following is the verification for the rule involved in the squaring of numbers ending with 5.

To simplify the discussion, we replace the series of digits which precede the final digit 5 with the symbol x. Notice, then, that the value of any number is always $10x + 5$. For example, the value of 2345 is $5 + (10 \times 234)$. Squaring the preceding altered representation of the number gives $(10x + 5)^2 = 100x^2 + 100x + 25$. This may be transformed to the expression $100x(x + 1) + 25$. In other words, we now have x multiplied by 1 greater than itself and that result multiplied by 100, to which we then add 25. Multiplying by 100 simply annexes two zeroes after the product $x(x + 1)$ and adding the final 25 just replaces those zeroes with the digits 25. The validity of the rule is thus verified.

CHAPTER 16

BOMBERS, BULLETS AND SHOCK WAVES

This is an investigation of the phenomenon generated as an object, such as a bullet or an airplane, travels through air at a speed greater than the speed of sound in that same air.

We will select the symbol V to represent the speed of the moving object and the symbol v to represent the speed of sound in air. As an aid in interpreting the two symbols, keep in mind that V, the larger symbol, represents the larger of the two speeds involved in the discussion. We will consider the situation that exists when a Stealth bomber travels along a horizontal line through air and at the speed of V. We know that whenever air is disturbed, as by a vibrating vocal cord, a spherical shell of compressed air, referred to as a compression wave, is emitted from the point of the disturbance. If any portion of this shell strikes a perceiving ear, a sound is heard and so we call the shell a sound wave. The diagram below illustrates a bomber moving along a straight line and disturbing the air at each point of its path.

DIAGRAM 1

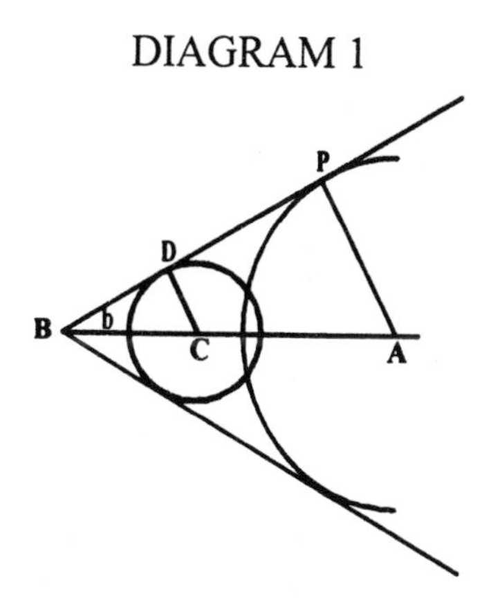

The symbol t is chosen to represent the variable time of the bomber's flight from an arbitrary fixed point A to point B. The circle whose center is at point A is a cross-sectional view of the spherical sound wave that was emitted as the bomber disturbed the air when previously passing point A. We choose a point P on the sound wave such that segment BP is tangent to the sound wave. This choice implies that triangle BPA is a right triangle. Since the time t that it takes the bomber to travel from point A to point B is the same as the time for the sound wave to travel from point A to point P, the length of segment AB is Vt and the length of segment AP is vt. As a consequence, $\sin b = v/V$.

If we choose any other point C through which the bomber has passed, a tangent from point B to point D on the sound wave which was emitted when the bomber passed point C will define another right triangle CDB. As before, the time that it took the bomber to travel from point C to point B is the same as the time for the sound to travel from point C to point D. So, as before, if we let T represent the new value of elapsed time, the length of segment CB = VT and the length of segment CD = vT. This implies that the sine of the angle at vertex B of triangle CDB is also v/V. So tangent BD is co-linear with tangent BP. This verifies that which our diagram has already suggested, viz., each sound wave emitted at point D of the bullet's path will be tangent to the straight line. On considering the three dimensional extension of these results, we see that every spherical sound wave emitted at the points along the path of the bomber will be tangent to the cone for which the diagram is a side view.

Since the family of shock waves that is tangential to the cone creates a super dense conical shell of air, it is possible with ultra high speed photography to photograph the shock wave. It is plausible that the refraction of light produced by the dense shell of air produces an image on the film. The angle displayed in the photograph and the known speed of sound enable us to determine the speed of the projectile producing the shock wave. In Diagram 1, for example, *b* was drawn as 30° and since $\sin 30^{\circ} = 0.5$, the bomber was traveling at twice the speed of sound.

If the bomber travels at exactly the speed of sound, the compression waves emitted at each point of its path will be arranged as shown in the following diagram. Every spherical shell of compressed air will be and will remain tangent to the nose of the bomber. In other words, all of the sound that proceeds in the direction of the bomber collects at its nose. If this condensed sound reaches an ear, there is a loud boom. This super dense air along the bomber's leading edges has been the source of many problems for aeronautical engineers.

DIAGRAM 2

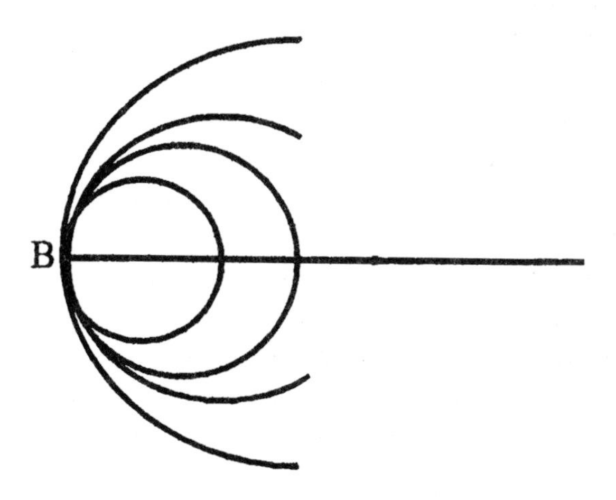

So far, it has been tacitly assumed that the entire body of involved air has been at a constant temperature. This has resulted in spherical compression shells. It is known, however, that air temperature decreases at the rate of approximately 3° F per 1000

feet of altitude for the lower atmosphere. Since the speed of sound varies as the air's absolute temperature, sound traveling upward from the bomber travels more slowly than the sound traveling downward. This condition will distort the shape of the sound waves. So we finish by determining the shape of these compression waves that are generated in this air of varying temperature.

PROBLEM: Given that: (1) The speed of sound in air varies as the absolute temperature of that air. (2) Air temperature decreases 3° F with each 1000 feet increase in altitude. (3) Sea level temperature is 92° F. This implies that the temperature of the air at the bomber's altitude will be 32° F 0° C. (4) The speed of sound in air at 0° C is approximately 1090 ft/sec. (5) 0° absolute is approximately equivalent to –460° F. What is the shape of the compression waves generated by a bomber flying horizontally at an altitude of 20,000 feet?

DIAGRAM 3

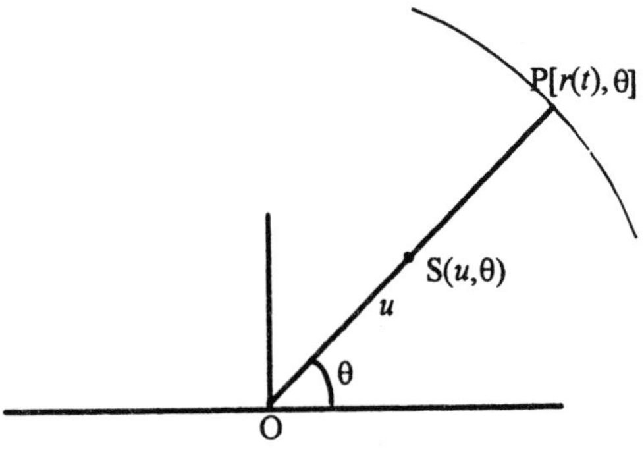

As we consider the problem, it soon becomes apparent that our investigation must rely on a quest for the polar coordinate equation of the compression wave that passes through point P at time t. With that in mind, we will refer to Diagram 3 and determine the polar equation of the family of compression waves that contains the member passing through point P.

The data that define our problem imply that the speed of sound at point S is

$$\text{sound speed} = 1090\sqrt{\frac{492 - 0.003r\sin\theta}{492}}.$$

Let us assume that point S is the point of intersection of segment OS and the compression wave generated *T* seconds ago, at point O. The rate at which *u*, the length of segment OS, changes with increasing *T* is the same as the speed of sound just mentioned. In other words,

$$\frac{du}{dT} = \frac{1090}{\sqrt{492}}\sqrt{492 - 0.003u\sin\theta}$$

In the interest of developing a general solution to the question at hand, we will let the general value *N* replace the specific value 1090 as the speed of sound at bomber's altitude. With this adjustment, the preceding equation produces the following integrals that then provide our solution.

$$\int_0^r \frac{1}{\sqrt{492 - 0.003u\sin\theta}}\,du = \int_0^t \frac{N}{\sqrt{492}}\,dT$$

Those who are conversant with the calculus will recognize that the solution is straightforward. We will allow them the option of obtaining this solution and proceed directly to our final result.

$$r = Nt - \frac{0.003N^2\sin\theta}{4(492)}t^2$$

The author recalls his academic exposure to limaçons and his adolescent doubt as to their utility. We now have verification of that utility. For each value of the parameter *t*, the locus just defined is a limaçon.

As indicated earlier, our current problem requires that $N = 1090$. We will use this to obtain a specific member of the family of compression waves. As an example, at $t = 2$ seconds after the compression wave is generated at point O, the polar coordinate equation of the expanded version of that wave would be:

$$r = 2180 - 7.77\sin\theta$$

As a result of the known range of values of $\sin\theta$, it is apparent that the minimum value of r occurs with $\theta = 90^{\circ}$ and the maximum value of r occurs with $\theta = -90^{\circ}$. So the temperature related distortion of the compression wave, from its normally spherical shape, is minimal. The distance from point O to the top of the 3-dimensional compression shell is only approximately 15.5 feet less than the distance to the bottom.

Remember that the shock wave we have been investigating was generated by an airplane which was traveling at a constant speed. The shock wave generated by a bullet traveling in air would be of a different shape. This is because the air resistance rapidly slows the bullet. In order to study the shock wave produced by such a bullet we must first obtain the relation that expresses s, the distance traveled by the bullet, as a function of elapsed time t. This we now do. In chapter 11 we discovered that a 150 grain bullet, traveling in air at 0° F and at a speed of 3000 ft/sec, encounters an air resistance force of approximately 100 gravities. Since, at high speeds, the air resistance varies as the square of the speed, we have force $F = kv^2$, where k is the air resistance constant. The 100 gravities of force implies that $F = 15{,}000$ grains when $v = 3000$ ft/sec.

Consequently k = 1/600 grain sec^2/ft^2. Now if we turn Newton's formula, $F = ma$, around so that $ma = F$, we have

$$\frac{150}{g}\frac{dv}{dt} = \frac{1}{600}v^2.$$

This enables us to develop the following relations defining the bullet's speed, v, and distance traveled, s, as functions of elapsed time t. Using $g = 32.15$ ft/sec^2, they are:

$$v = \frac{2800}{t + 0.9331} \quad \text{and} \quad s = 2800[\ln(t + 0.9331) + \ln(1.0717)]$$

Below is a scale drawing which displays the sound wave spheres emitted at the instant the bullet left the muzzle at point M, at the end of 0.2 seconds, and at the end of 0.4 seconds. Since the bullet is decreasing in speed, the three dimensional supersonic shock wave is not a cone. The side view below shows how its shape differs.

DIAGRAM 4

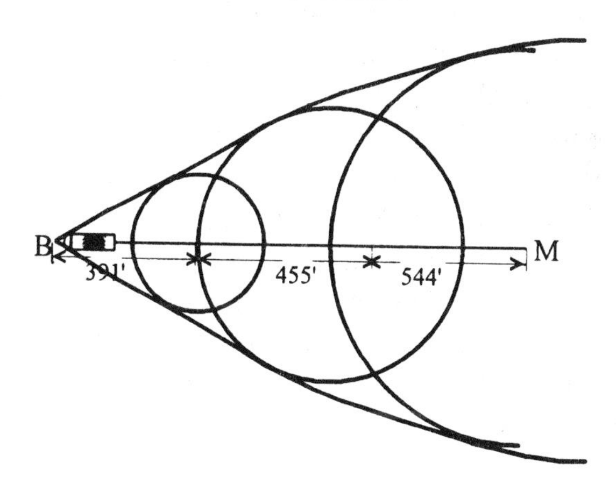

A Morsel (food for thought)

Ah, the magic of numbers! The three pairs of statements that follow, display a sample of that magic.

$1 + 2 = 3$
$1^3 + 2^3 = 9$

$1 + 2 + 3 = 6$
$1^3 + 2^3 + 3^3 = 36$

$1 + 2 + 3 + 4 = 10$
$1^3 + 2^3 + 3^3 + 4^3 = 100$

In time it becomes apparent that, in each pair of equations, the second sum is the square of the first sum. This makes us wonder whether or not this relation is true in general. A well known and easily proved fact is that the sum of the first n integers is equal to $\frac{n(n+1)}{2}$. So can it be true that the sum of the cubes of the first n integers is always equal to $\left[\frac{n(n+1)}{2}\right]^2$? The answer is yes.

A method favored by most mathematicians for verifying such theorems involving integers is the process of mathematical induction.

A procedure for deriving results for similar sums of integral powers of integers will be presented as the next morsel.

CHAPTER 17

A Lunar Look

The procedure for determining the time span between successive full moons is less complicated than the procedure for determining the period of the moon's revolution about the earth. The investigation involved here will demonstrate that the time between full moons is dependent on this period of revolution. The latter topic will be discussed in Chapter 21. The time between full moons is greater than the moon's period of revolution. This information is necessary for the arrangement of the moon in the diagram which follows.

Due to the non-circularity of the orbits of moon and earth, we can expect our finding to be only an average length of time between full moons. The situation which exists in the arrangement of sun, earth and moon at successive full moons is illustrated in the diagram.

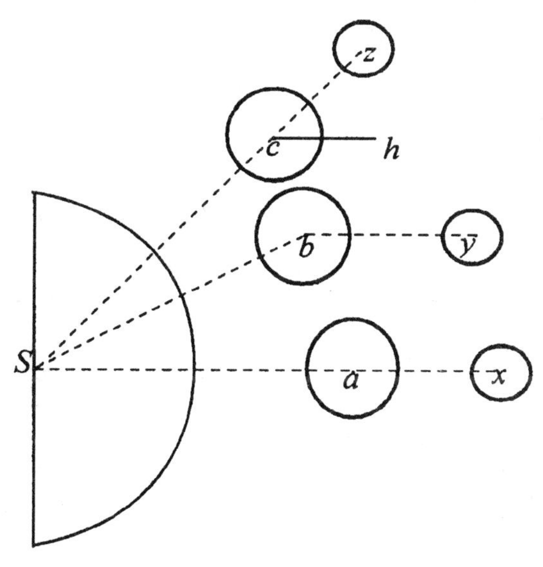

Of course the diagram does not depict the sun, earth and moon to proper scale. The diameter of the sun is more than 100 times that of earth, and earth's diameter is nearly 4 times that of the moon. Also, the distance from earth to sun is nearly 400 times the distance from earth to moon.

If we assume that we are somewhere in space looking down on the earth's north pole, the revolution of both earth and moon in their orbits and the rotation of both earth and moon about their axes are all in a counterclockwise direction. The diagram indicates the positions of sun, earth and moon at initial full moon (bottom), at one moon revolution (next above) and at the succeeding full moon (at top). Segments Sa, yb, and hc are parallel.

Since the diagram is limited to two dimensions, the center of the moon appears to be in the plane of the earth's orbit. Very seldom is this the case, but this will be irrelevant to the discussion. The plane of the moon's orbit is tilted with respect to that of the diagram, which is also the plane of the earth's orbit. The center of the moon passes through the plane of earth's orbit twice during each moon revolution. Whenever one of those occasions coincides with the time of full moon, there will be a total eclipse of the moon somewhere on earth. As the diagram indicates the moon will be in the shadow of the earth – that is, it will be eclipsed only at or near the time of full moon.

We choose the following symbols: E = number of days for one revolution of earth about the sun; F = number of days between successive full moons; M = number of days for one revolution of moon about earth.

As the moon travels from point x to point y, it revolves through 360° in M days. As the moon travels from point x to point z, it revolves through $\frac{F}{M}360^o$ in F days. As the earth travels from point a to point c it revolves through $\frac{F}{E}360^o$ in F days. Now we note that $\angle hcz$ is the difference between the first two moon revolutions angles mentioned above, so

$$\angle hcz = \frac{F}{M}360^o - 360^o.$$

Also $\angle aSc = \frac{F}{E}360^o$.

But $\angle hcz = \angle aSc$, so we see that

$$\frac{F}{M}360^o - 360^o = \frac{F}{E}360^o$$

After dividing by 360°, clearing the fractions, transposing and factoring, this becomes $F(E - M) = EM$. The resulting value of F is:

$$F = \frac{EM}{E - M}$$

In the event that the rolès of F and M are reversed, the preceding formula converts to the following equivalent.

$$M = \frac{EF}{E + F}$$

Given the time for one moon revolution, 27.322 days, and the earth's year, 365.24 days, the formula's result for the average time between successive full moons is 29.531 days. According to the almanac, the average time for 11 successive full moons during 1997 was 29.50 days. So the effect of the sun's gravitation and the non-circularity of the moon's orbit alters the period by only approximately 0.03 days.

A Morsel (food for thought)

Presented here is the unique procedure that enables us to develop all expressions providing the sum of the first n integers each raised to the same positive integral power. The technique will be illustrated as we develop the expression which provides the sum of the squares of the first n integers.

Applying the algebra of summations, we first generate the following identity:

$$\sum_{i=1}^{n}(i+1)^3 = \sum_{i=1}^{n} i^3 + 3\sum_{i=1}^{n} i^2 + 3\sum_{i=1}^{n} i + \sum_{i=1}^{n} 1$$

Initially we recall that $\sum_{i=1}^{n} i = \frac{n(n+1)}{2}$ and $\sum_{i=1}^{n} 1 = n$. We then use these relations to rearrange the preceding identity in the following manner:

$$\sum_{i=1}^{n}(i+1)^3 - \sum_{i=1}^{n} i^3 = 3\sum_{i=1}^{n} i^2 + 3\frac{n(n+1)}{2} + n$$

All that remains on the left side of the equation after the indicated subtraction is $(n+1)^3 - 1$. You may now verify that the proper algebraic manipulation produces the result we have hlonged for:

$$\sum_{i=1}^{n} i^2 = \frac{n(n+1)(2n+1)}{6}.$$

Thus we have the formula for the sum of the first n squares.

By following this exemplified pattern, we can then determine the sum of the first n cubes, then the sum of the first n fourth powers, etc. The trick is to start with $\sum (i+1)^{p+1}$ if the sum of the first n integers each to the p power is desired.

CHAPTER 18

SOME TERRESTRIAL MEASUREMENTS

The next three chapters present and apply those elements of spherical geometry and trigonometry necessary for terrestrial and solar measurements. Exercises included in the following chapters will introduce us to some of the concepts and also provide worthwhile information.

Before considering the topics involved in this chapter, it is necessary to highlight several facts concerning the geometry and trigonometry of a sphere. They will prove critical to the understanding of the concepts to be discussed.

A great circle of a sphere is a circle formed by the intersection of the sphere with any plane passing through its center. Every great circle divides the sphere into two hemispheres, and no circle on that sphere can be larger. If the earth were a perfect sphere, its equator and all meridians would be great circles. On the other hand those circles on earth called parallels, circles of equal latitude, are not great circles. Such circles are called small circles. As will be demonstrated later, the radius of a parallel of latitude is equal to the cosine of that latitude times the radius of the earth.

Relevant to this discussion will be a fascinating theorem from solid geometry. It is applicable to the geometric geography which is to be discussed.

Theorem

If a slice of uniform thickness h is removed from a sphere, the surface area of the sphere contained on the slice depends only on the radius of the sphere and the thickness of the slice. It is not dependent on the slice's position in the sphere.

This theorem is illustrated by the following formula for the lateral area of any spherical slice of thickness h which is removed from a sphere of radius r: $A = 2\pi rh$.

Suppose that a sphere is inside a cylindrical container whose height and diameter are the same as the diameter of the sphere. An analysis of the preceding formula for the area of a spherical slice will provide this surprising result. A horizontal slice of uniform thickness will remove equal areas from both the sphere and cylinder. As a corollary, the lateral area of the cylindrical container is the same as the area of the sphere.

A spherical triangle is any portion of a sphere's surface that is bounded by arcs of three distinct great circles. An example of such appears in conjunction with the solution of a problem soon to be presented. The lengths of the sides of a spherical triangle are measured in degrees or radians of arc. The angle at a vertex of the triangle is defined to be the angle between the two lines through the vertex that are tangent to the respective sides. Unlike a plane triangle, the sum of the angles of a spherical triangle is not 180°. Instead, the sum may have any measure between 180° and 540°. As the sum of the angles, which we will represent as S, approaches 180°, the sides of the triangle reduce to a point or else they approach a common great circle arc. As the sum S approaches 540° the sides of the triangle will approach a complete great circle, in which case the area of the triangle will approach the area of a hemisphere.

The area of a spherical triangle depends only on this sum, S, and the radius of the sphere. If degree measure is used, this area is given by the formula $A = (\pi/180^\circ)(S - 180^\circ)r^2$, where r is the radius of the sphere. If radian measure is used, the area function simplifies to $(S - \pi)r^2$.

In all following discussions, whenever the geometry of the earth is involved, we will use R to represent its radius. This presents a problem, since the earth is not a perfect sphere, but rather an ellipsoid, sometimes referred to as an oblate spheroid. The earth's equatorial radius is 3963.2 statute miles and its polar radius is 3949.9 statute miles. We will also be referring to a nautical mile, the length of one minute of arc of a great circle of earth, so its length depends on the radius measurement. Although it makes us prone to slight inaccuracies, we opt for the convenience of consistency and will use 6080 feet as the length of a nautical mile. This requires that we also adopt 3958.6 statute miles as the radius of earth.

Suppose that a huge square ice rink, 2 miles on a side, is constructed. Since the rink is a portion of the earth's surface, it forms a spherical instead of a plane square. Therefore, its area will slightly exceed 4 square miles. In order to compute the rink's exact area, we will construct a diagonal of the rectangle and divide the rink into two equal spherical triangles. But, in order to determine the angles of these triangles, we need to learn a few fundamentals of spherical trigonometry. It is now necessary to investigate the relations involving the parts of the spherical triangle displayed below.

DIAGRAM 1

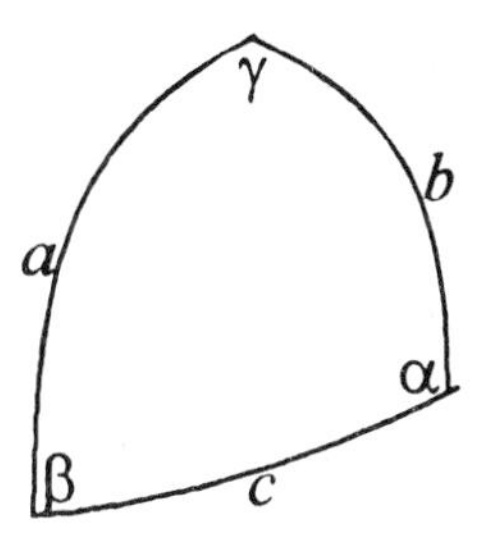

Most spherical triangle problems can be solved with the aid of the following two formulas.

1. Sine law: $\dfrac{\sin a}{\sin \alpha} = \dfrac{\sin b}{\sin \beta} = \dfrac{\sin c}{\sin \gamma}$
2. Cosine law: $\cos a = \cos b \cos c + \sin b \sin c \cos \alpha$

The cosine law has many more applications than the sine law.

Before studying this general spherical triangle, we will first analyze the right triangle produced when $\gamma = 90°$.

To thoroughly analyze a right spherical triangle requires ten separate equations, but fortunately the mathematician, Napier, created a device which eliminates the necessity for memorization. In Diagram 2 below a right spherical triangle and Napier's memorization device are displayed.

DIAGRAM 2

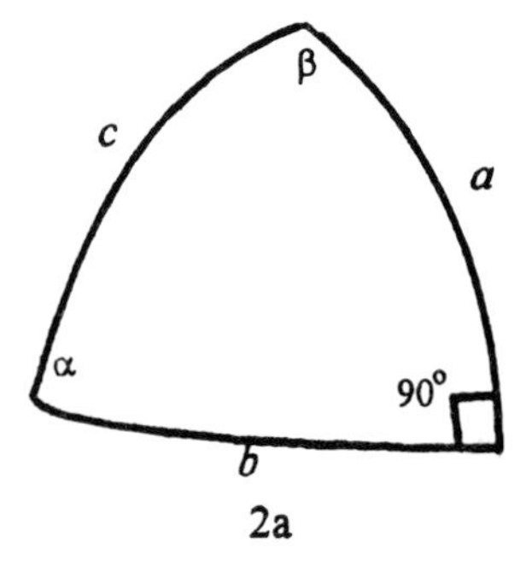

2a

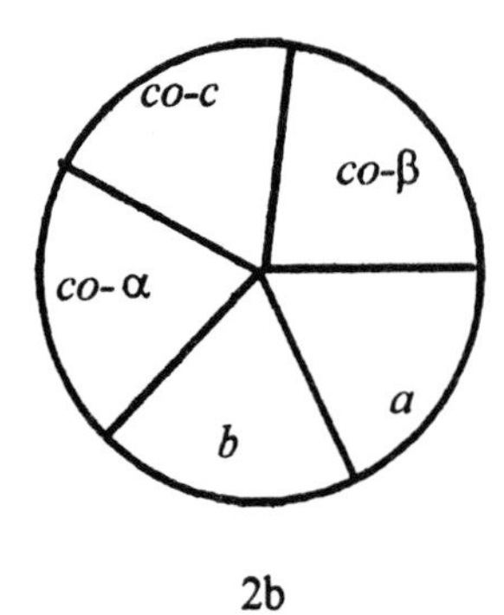

2b

Notice that, in the circle, all parts other than the sides that form the right angle have the prefix *co-*. With the right angle omitted, the order of parts in the circle is the same as in the triangle. Referring to the sectored circle, all ten relationships involving the parts of a right spherical triangle are provided by the following three rules.

1. The sine of any part is equal to the product of the cosines of its two opposite parts.

2. The sine of any part is equal to the product of the tangents of its two adjacent parts.

3. Wherever the prefix *co-* appears, the co-function of the function provided by the first two rules must be used.

EXAMPLE 1. Rule 1 tells us that $\sin(co\text{-}c) = \cos a \cos b$.
Rule 3 converts this to the equation

$$\cos c = \cos a \cos b.$$

EXAMPLE 2. Rule 2 tells us that $\sin a = \tan(co\text{-}\beta)\tan b$.
Rule 3 converts this to the equation

$$\sin a = \cot \beta \tan b.$$

With this latter formula, we are now able to solve for the area of the previously mentioned ice rink.

DIAGRAM 3

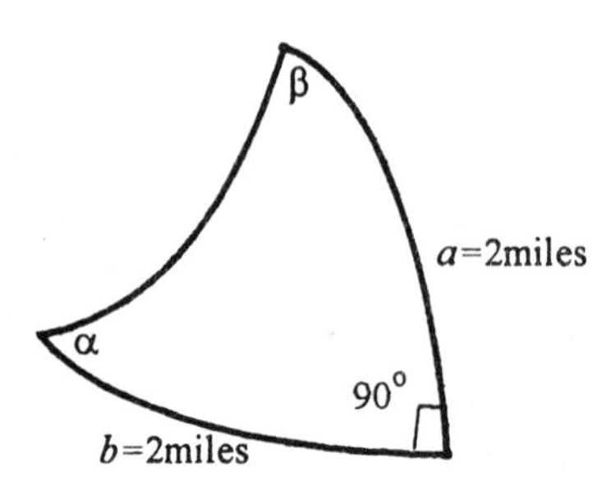

In Diagram 3, we are given that $\angle\alpha = \angle\beta$, so we solve for β which appears in the formula already obtained. This gives $\cot\beta = \dfrac{\sin a}{\tan b}$. Since $a = b$, the application of trig identities converts the latter equation to $\tan\beta = \dfrac{1}{\cos b}$.

Next we must determine the measure of a and b, which are equal arcs of a great circle on earth. In order to obtain the degree measure of a and b, we must multiply $180^\circ/\pi$ by the number of radians in arc a. The number of radians in any arc is the length of the arc divided by the radius of its circle. Using R = 3960 mi., we obtain the terrestrial arc a, 2/R = 2/3960. As a result, this produces $a = b = 0.02894749634^\circ$. Then, on solving for β we have $\alpha = \beta = 45.00000365^\circ$. Using $S = (2\alpha + 90^\circ)$, and substituting this in our formula for the area of a spherical triangle, we obtain A = 2.000000087 square miles. Hence the area of the square rink would be twice this number or 4.000000174 square miles. Therefore, the area of a spherical square 2 miles on a side has an area of 4 square miles plus approximately 5 square feet.

Of course, not all terrestrial area problems involve spherical triangles. The states of Colorado and Wyoming are interesting in that they are both 4° from south to north and 7° from west to east. If you are tempted to assume them to have equal areas, don't. Since the meridian boundaries converge as they proceed northward, Colorado's average west to east dimension is greater than Wyoming's. Since meridians are great circles, the south to north distance for each state is $4 \times 69.09 = 276.4$ statute miles. In order to determine the length of the northern and southern boundaries, we refer to Diagram 4 below.

DIAGRAM 4

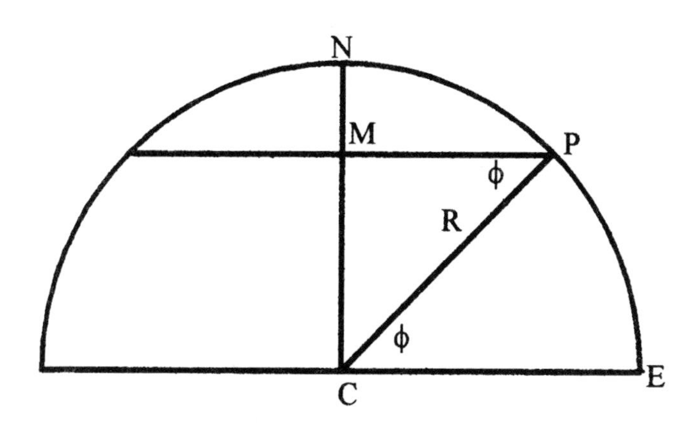

In the diagram, PM is the radius of the parallel of latitude through point P.

Notice that $\angle$ MPC = $\angle$ PCE = **latitude of point P** = ϕ. Consequently PM, the radius of the circular parallel of latitude, is given by

$$PM = R \cos \phi.$$

The distance from the center of the earth to the center of the parallel circle is

$$CM = CP \times \sin \phi = R \times (\text{sine of the latitude of point P}).$$

Using the first of the preceding results, it follows that the length of 1° on a parallel of latitude is equal to the length of a degree of earth's great circle times the cosine of that latitude. Hence, the length of the southern boundary of Colorado would be $7 \times 69.09 \cos 37^{\circ} = 386.2$ statute miles. The northern boundary of Colorado, obtained similarly, is 365.0 statute miles; the southern and northern boundaries of Wyoming are, respectively, 365.0 and 342.0 statute miles.

Since two of the boundaries of each state are not great circle arcs, the methods used to determine the ice rink area cannot be applied when computing the areas of the states. Happily, this simplifies the problem. Focusing on Colorado, its 7° length implies that its area is 7/360 times the area of the spherical slice of earth's surface between the 37th and 41st parallels. We previously learned that this latter area is $2\pi Rh$, where h = the thickness of the slice. The thickness would be the distance of the top of the slice from the earth's center minus the distance of the bottom of the slice from the earth's center. We find that these two distances are $R \sin 41^{\circ}$ and $R \sin 37^{\circ}$. Hence, $h = R(\sin 41^{\circ} - \sin 37^{\circ})$. Combining the several relevant relations, and using R = 3958.6 mi., we multiply 7/360 by the area between the parallels at the northern and southern boundaries. The result for the area of Colorado is 103,850 square miles. This of course assumes that Colorado is a portion of a perfect sphere, so the actual area will be slightly different. The same procedure provides 97,730 square miles as the area of Wyoming.

Two other questions which may be of interest are:

1. How much shorter is the great circle distance between the ends of Colorado's northern border than is the distance measured along that border?

2. What is the maximum distance between this great circle path and the small circle which forms the state's northern border?

We are unable to solve these problems with the aid of the material here presented, but the tools developed in the next chapter will enable us to verify the answers listed below.

1. The great circle distance between the ends of the northern border is 542 feet less than the length of that border.

2. The maximum departure of the preceding great circle arc from the northern border is 3.66 miles.

A MORSEL (food for thought)

From Golden High School

Rules For Teachers in 1872

1. Teachers will each day fill lamps, clean chimneys.
2. Each teacher will bring a bucket of water and scuttle of coal for the day's session.
3. Make your pens carefully. You may whittle nibs to the individual taste of pupils.
4. Men teachers may take one evening each week for courting purposes, or two evenings a week if they go to church regularly.
5. After ten hours in school, the teachers may spend the remaining time reading the bible or other good works.
6. Women teachers who marry or engage in unseemly conduct will be dismissed.
7. Every teacher should lay aside from each pay a goodly sum of his earnings for his benefit during his declining years so that he will not become a burden to society.
8. Any teacher who smokes, uses liquor in any form, frequents pool or public halls, or gets shaved in a barber shop will give good reason to suspect his worth, intention, integrity and honesty.
9. The teacher who performs his labor faithfully and without fault for five years will be given an increase of twenty-five cents per week in his pay providing the Board of Education approves.

Golden, Colorado

CHAPTER 19

Some Terrestrial Navigation

This initial section explaining the rudiments of navigation will involve some plane geometry principles that enable us to use the sun and the North Star. We will learn the power of solar observations and how the North Star, Polaris, can be used to determine latitude. We will also learn how to apply some of the spherical trigonometry presented in the preceding chapter to some basic terrestrial navigation. Toward that end, we need the following diagrams.

DIAGRAM 1

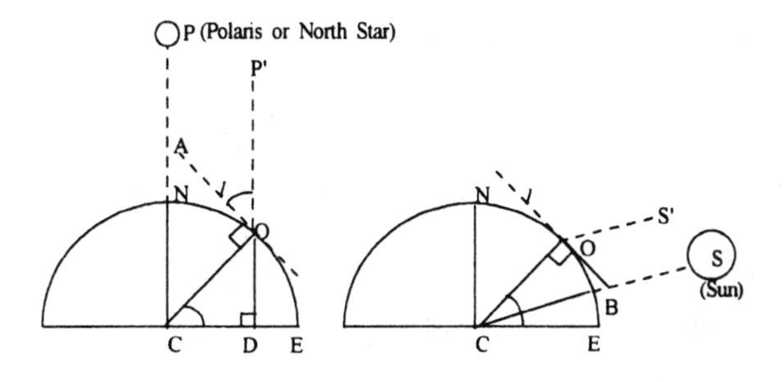

Diagram 1a Diagram 1b

Diagrams 1a and 1b both depict the earth's northern hemisphere. Point C is the center of the earth, point E is a point on the earth's equator and N is the North Pole. An observer is at point O and the observer's latitude is defined to be the measure of $\angle$ECO. Line l is tangent to the observer's meridian at O.

Consider now Diagram 1a. Polaris (the North Star) is at P. The line through P′ and O is parallel to the segment CP and intersects segment CE at D. Because of the great distance from the observer to Polaris we can consider segment OP′ to be the line of sight from the observer to Polaris.

Plane geometry teaches us that:

$$\angle AOP' + 90^\circ + \angle COD = 180^\circ \text{ and}$$
$$\angle OCD + 90^\circ + \angle COD = 180^\circ.$$

Therefore

$$\angle AOP' = \angle OCD.$$

Happily, the first angle is the angle of elevation, called the altitude, of the North Star while the latter angle is the latitude of the observer. Thus we have learned that:

latitude of observer = angle which observer sees as the altitude of the North Star.

As often happens, these measurements are subject to slight adjustments. Polaris is not exactly on the extension of the earth's axis, but is sufficiently close for our purposes. Then too, there is a small correction for atmospheric refraction of light entering from space. However, this refraction is small. The rays coming from a star on the horizon are refracted approximately 36 minutes. But when the star has risen 10°, the refraction angle is only 2.5 minutes. Hence this correction will also be ignored here.

In Diagram 1b we consider the situation that exists at the time of the observer's solar noon. Position S will be referred to as the position of the vertical rays of the sun. Line *l* intersects line segment CS at B. Segment OS′ is parallel to segment CS. Again, because of the great distance involved, we may consider segment OS′ to be the line of sight from the observer to the sun. $\angle ECO$ is the latitude of the observer. $\angle BOS'$ is the solar noon angle of elevation of the sun. We will usually use the term altitude instead of angle of elevation. $\angle ECS$ is the declination of the sun. (Knowing the day of the year, this angle can be found in navigation charts.)

Since OS′ and CS are parallel, $\angle SCO + \angle COS' = 180^\circ$.
Also, since BO is tangent to the circle at point O, $\angle COB = 90^\circ$.
So $\angle BOS' + \angle SCO = 90^\circ$. Therefore,

$$\begin{aligned}\angle BOS' &= 90^\circ - \angle SCO \\ &= 90^\circ - (\angle ECO - \angle ECS) \\ &= 90^\circ + \angle ECS - \angle ECO.\end{aligned}$$

Interpreting these angles we know that at solar noon:

Altitude of sun = 90° + declination of sun
− latitude of observer.

This gives us a very important principle:
The altitude of the sun is equal to 90° minus the measure of the arc joining the position of the observer and the position of the vertical rays of the sun.

If we wish to focus on the observer's latitude, we simply rearrange the preceding equation and obtain

Latitude of observer = 90° + declination of sun
− sun's altitude at solar noon.

This then, is a second procedure for determining an observer's latitude.

Of course, in order to apply these findings we need to know when solar noon occurs and what the declination is. These topics will be discussed later.

We will now consider an interesting and practical application of the cosine law presented in the preceding chapter. This law enables us to develop the procedure for determining the distance between any two points on the surface of the earth. We will use the law to obtain the great circle distance between two points whose latitudes and longitudes are known.

In Diagram 2, consider the spherical triangle with vertices at points B, N, and the North Pole, P. The ordered pairs (u, v) and (x,y) will represent, in that order, the latitudes and longitudes of B and N.

DIAGRAM 2

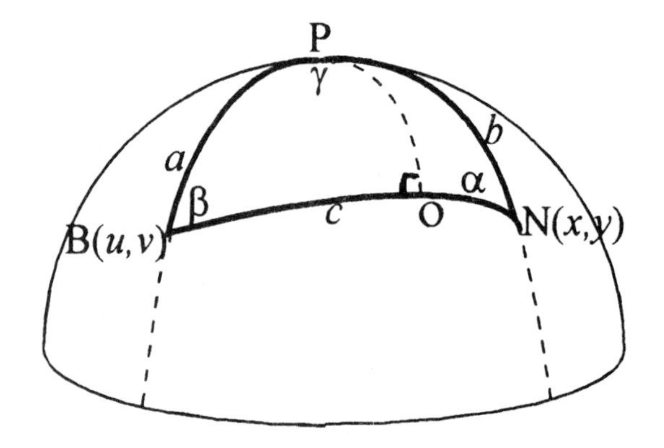

Notice that $\angle \gamma = v - y$, the difference in the longitudes of B and N. Also, a and b are both 90° minus the respective latitudes of B and N: $a = 90° - u$ and $b = 90° - x$. Now that three parts of the spherical triangle are known, the cosine law will enable us to determine c, the distance between B and N. To refresh the reader's memory, we reproduce the cosine law.

$$\cos c = \cos a \cos b + \sin a \sin b \cos \gamma$$

Since $a = 90^{\circ} - u$, any trigonometric function of a will be replaced by its cofunction of u. Similarly, any function of b will be replaced by its cofunction of x. Substituting the given variables into the cosine law, we learn that:

$$\cos c = \sin u \sin x + \cos u \cos x \cos(v - y)$$

To accommodate the situation when B and N are in different hemispheres, we adopt the convention that latitudes are positive in the northern hemisphere and longitudes are positive in the western hemisphere. Hence, the latitude of a point in the southern hemisphere must be assigned a negative value and a longitude in the eastern hemisphere must also be assigned a negative value. Now we are in position to determine c, the arc between B and N.

Assume that N is the position of New York, latitude north 40.75° and longitude west 74.00°, while B is the position of Beijing with latitude north 39.93° and longitude east 116.40°. We are now ready to find the distance between these two cities. Since the latter longitude is negative (B is in the eastern hemisphere), the cosine law provides the following relation:

$$\cos c = \sin 40.75^{\circ} \sin 39.93^{\circ} + \cos 40.75^{\circ} \cos 39.93^{\circ} \cos 169.60^{\circ}$$

On solving for c, we find that $c = 98.77^{\circ}$.

Now we recall that a nautical mile is the length of 1 minute of arc of earth's great circle and we have adopted 6080 feet as its length. As a result, 1 nautical mile = 1.1515 statute miles. Therefore 1° of arc of a terrestrial great circle has a length of 69.09 statute miles. After multiplying the number of degrees in arc c by 69.09, we determine the distance from New York to Beijing to be 6824 statute miles.

Next we consider an airplane that is capable of flying nonstop from New York to Beijing. Now that we have obtained all three sides of the involved triangle, it is possible to use the same cosine law to obtain the course of the plane as it leaves New York. It would be the $\angle \alpha$. Substituting the known lengths of the triangle's three sides in the cosine law provides the equation which determines α.

$$\cos 50.07^{\circ} = \cos 49.25^{\circ} \cos 98.77^{\circ} + \sin 49.25^{\circ} \sin 98.77^{\circ} \cos \alpha$$

Solving for α provides the result $\alpha = 8.052^{\circ}$. In other words, the initial course of the plane leaving New York is only 8.052° west of due north.

Finally, consider an arc from the North Pole, P, and perpendicular to the airplane's path at point O. Then use the resulting right spherical triangle NOP. The distance OP is the minimum distance from the North Pole to the airplane's path. Therefore, the position of point O is the airplane's closest position to the pole. Determining the length of OP and point O's latitude and longitude would require the application of Napier's rules for solving the right spherical triangle, NOP. These rules were stated in the preceding chapter on page 82.

Determining the latitude and longitude of point O is left as an exercise. The answers are on the next page.

ANSWERS:

The distance OP is 6.09° or 420.8 statute miles, making the latitude of point O, 83.91°.

∠NPO is 84.72°, which when added to the longitude of New York makes the longitude of point O, 158.72°.

A Morsel (food for thought)

We all have been exposed to tales of deep sea exploits and of the tremendous pressures there encountered. Some of us may have had difficulty fathoming the origin of the large numbers mentioned. Here is the explanation.

Suppose that a container in the form of a cube with 1 foot dimensions is filled with fresh water. The standard weight of 1 cubic foot of fresh water is 62.43 pounds, so this is the force pushing on the entire bottom of the cube. Consequently, each of the 144 square inches comprising the bottom of the cube must support a force of $\frac{62.43}{144} = 0.434$ pounds. In other words, at a depth of 1 foot, fresh water exerts a pressure of 0.434 lb/in^2. Since water is virtually incompressible, the pressure increases this same amount for each additional foot of depth. If sea water is involved, it is necessary to multiply by 1.025, to adjust for the greater density of sea water. So pressure in sea water increases at the rate of 0.444 lb/in^2 for each foot of depth. That's why the water pressure at a depth of 10,000 feet beneath the surface of the ocean is 4440 pounds on each square inch.

CHAPTER 20

FUN WITH THE SUN

Now that we have been exposed to solar observations and the power of spherical trigonometry, it is just a matter of time before we gaze at the sun and wonder if that knowledge might assist us to predict a future sun position or to solve some of life's other significant problems. The time in "matter of time" is **now**. And the significant problems include calculating the time of sunrise and perhaps finding the direction of the shadow of that notorious groundhog.

Toward that end, we will analyze the terrestrial spherical triangle shown in Diagram 1.

DIAGRAM 1

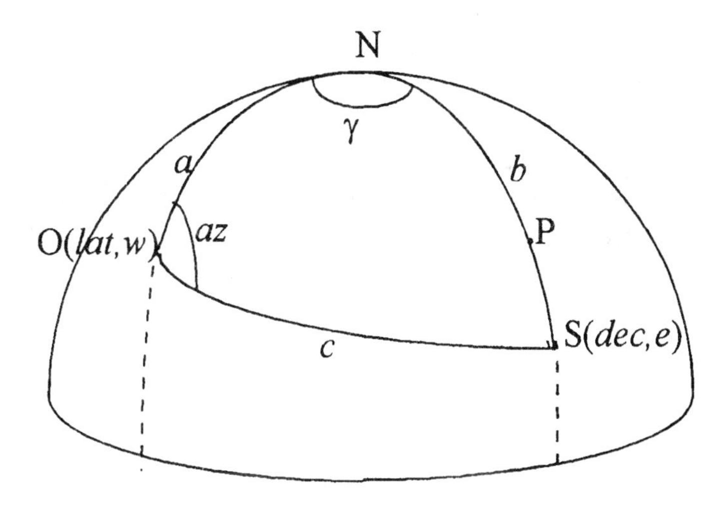

In this spherical triangle, vertex N represents the North Pole, vertex O represents the position of the observer and vertex S represents the position of the vertical rays of the sun at the time of the observation. The coordinates of O are *lat* and *w*: *Lat* represents the latitude of the observer and *w* represents the corresponding longitude. The coordinates of S are *dec* and *e*: *Dec* represents the latitude of the sun's vertical rays – called the sun's declination – and *e*, represents the corresponding longitude. The longitude symbols *w* and *e* were chosen for their west-east relationship during our diagrammed morning observation. The symbol *alt* will represent the angle of elevation of the observed sun, called the altitude of the sun. The symbol *az* will represent the azimuth of the sun. The azimuth is defined to be the angle between the true north line through point O and the vertical plane containing both point O and the sun's center. Since the coordinate *lat* is defined to be the measure of the dotted arc which extends arc NO to the equator, side $a = 90^{\circ} - lat$. An identical argument implies that side $b = 90^{\circ} - dec$. Now we will develop a corresponding result for side *c*.

As we developed the formula for determining the sun's altitude in the preceding chapter, it was stated that a very important principle that had just been developed was: ***The altitude of the sun is equal to 90° minus the measure of the arc joining the position of the observer and the position of the vertical rays of the sun.*** From this relation we may infer the important equivalent relation: The measure of the arc joining the position of the observer and the position of the vertical rays of the sun is equal to 90° minus the altitude of the sun. But our diagram indicates that the symbol *c* represents this measure of the arc joining the observer and the position of the vertical rays, point S. Thus we have verified the very important result: $c = 90^{\circ} - alt$.

All references to time in this investigation must be measured from the moment of the observer's solar noon. By definition, solar noon at a given position occurs at the moment when the sun's vertical rays intersect the meridian passing through that position. In our diagram, it is solar noon at point S and at every point on the meridian containing point S.

It might be worthwhile to ponder the reason for the seasonal change in the sun's declination. If the earth's axis were perpendicular to the plane of the earth's orbit, the earth's equator would always lie in this orbital plane. In this case a line joining the centers of earth and sun would always pass through the equator. In other words, the sun's vertical rays would always lie on the equator and the earth would have no seasons. The equatorial region would be hotter than it is and the polar regions would be colder. Fortunately for us the earth's axis is tilted, the axis forming a 23.45 degree angle with a line perpendicular to the orbital plane. Now consider a plane which at all times contains the earth's axis and the sun's center. Twice a year this plane will be perpendicular to the earth's orbital plane. At these times the latitude of the sun's vertical rays will be at maximum north or south latitudes. At one of these times, when our North Pole is in sunlight, the sun's declination has the maximum value 23.45 degrees north and we have the northern hemisphere's summer solstice. The southern hemisphere's summer solstice, equivalent to the northern hemisphere's winter solstice, may be explained similarly.

Since the sun's vertical rays make a 360° circuit of our globe in 24 hours, the longitude of those vertical rays must increase at the rate of 15 degrees per hour. This information has relevance to the evaluation of the angle γ. γ, shown in Diagram 1, measures the difference in the longitudes of points S and O:

γ = **$15^\circ \times$ (number of hours from time of solar observation to solar noon at this point of observation).**

For example, if a solar observation is made at 9:40 solar time, or 7/3 hours before the observers solar noon, γ will be $7/3 \times 15^\circ$, or

35°. Note: Since γ is a function of time only, we will call it the "*time angle*".

Since $a = 90° - lat$, $b = 90° - dec$ and $c = 90° - alt$, every application of the trigonometric formulas will be subject to the following conversions. Each function of a will be replaced by its cofunction of *lat*. Each function of b will be replaced by its cofunction of *dec*. Finally, each function of c will be replaced by its cofunction of *alt*. For example, $\cos a = \sin lat$ and $\tan b = \cot dec$.

Both declination and solar noon are obtainable from an astronomical almanac, but both declination and solar noon can be closely approximated by methods that will be presented in the latter part of this chapter. If solar noon is known at any meridian, the 15 degrees per hour principle enables us to determine solar noon at all longitudes. Atmospheric refraction of light will be ignored in our calculations, although it too is obtainable from astronomical tables.

We are now prepared to discover the procedure for determining both the sun's altitude and azimuth at any given solar time. To accomplish this we again rely on the cosine law:

$\cos c = \cos a \cos b + \sin a \sin b \cos \gamma$, which converts to:

$$\mathbf{\sin \mathit{alt} = \sin \mathit{lat} \sin \mathit{dec} + \cos \mathit{lat} \cos \mathit{dec} \cos(\mathit{time\ angle})}$$

Now that we have the formula for determining the altitude of the sun it is also possible to determine our other objective, the sun's azimuth. To accomplish this, we apply the previously mentioned sine law and learn that $\frac{\sin az}{\sin b} = \frac{\sin \gamma}{\sin c}$, which is transformable into $\dfrac{\sin az}{\cos dec} = \dfrac{\sin(time\ angle)}{\cos alt}$. From this emerges the formula

sought: $\mathbf{\sin az = \dfrac{\sin(\mathit{time\ angle})\cos \mathit{dec}}{\cos \mathit{alt}}}$

Also, if just the latitude, declination and sun's altitude are known, the cosine law informs us that:

$$\cos az = \frac{\sin dec - \sin lat \sin alt}{\cos lat \cos alt}$$

All of which has brought us to a juncture. So at this juncture we will apply our newly acquired knowledge.

Suppose that a vertical flagpole stands on a horizontal surface. We are given that the pole's height is 20 feet, the pole's latitude is 40° north, solar noon is at 12:10 P.M. and declination is 10°15', or 10.25°, south. Where will the tip of the flagpole's shadow be at 10:30 A.M.?

DIAGRAM 2

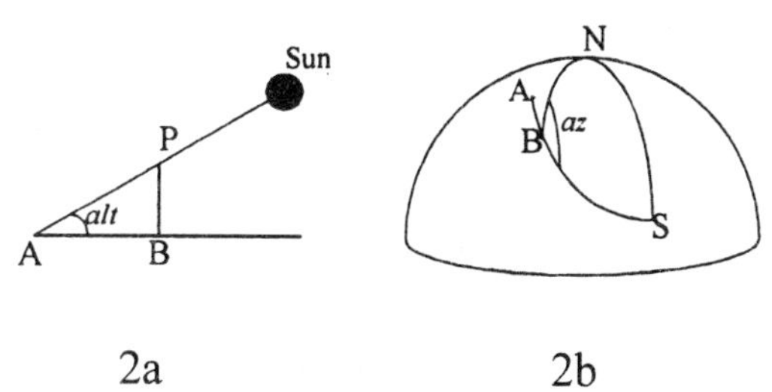

2a 2b

Locating the position of the shadow's tip requires the determination of both the sun's altitude and azimuth, illustrated in Diagram 2a and 2b respectively.

In Diagram 2a, point P is the top of the pole, point B is its base, point A is the tip of the shadow and *alt* is the sun's altitude.

Since the time of observation, 10:30 A.M., is 1 hour and 40 minutes or 5/3 hours prior to the observer's solar noon, the *time angle* is $5/3 \times 15^\circ$ or 25°. The declination is south, so it is -10.25°. With this information we apply the relevant cosine law and, after substituting the cofunctions, we obtain

$$\sin alt = \sin 40^\circ \sin(-10.25^\circ) + \cos 40^\circ \cos(-10.25^\circ) \cos 25^\circ.$$

As a result, the sun's altitude turns out to be 34.67°. The length of the shadow is the length of segment AB.

$$AB = \cot alt = 20 \cot 34.67^\circ = 28.92 \text{ feet.}$$

In Diagram 2b, B is again the base of the pole and point A is the tip of its shadow. S is the position of the vertical rays of the sun and *az* is the azimuth of the sun. Now applying the formula recently developed for determining the azimuth, we obtain

$$\sin az = \frac{\sin 25^\circ \cos(-10.25^\circ)}{\cos 34.67^\circ}.$$ Thus, $\sin az = 0.50566$.

Since the declination is negative, we know that the date of the observation must be after the autumnal equinox and prior to the vernal equinox. During this period the sun is always to the south of the given observer's latitude, so the azimuth must be an angle between 90° and 180°. Therefore the azimuth is 149.62°. As a result the line of the shadow, AB, lies 30.38° to the west of north.

Readers with inquiring minds may find it interesting that our sun's azimuth and altitude formulas apply even after dark. They are invited to verify the following scenario. We confidently, practically, guarantee its accuracy. Suppose that at 40° north latitude a really powerful laser has drilled an unobstructed hole east of north through the earth, forming an angle of 13.95° downward from a horizontal plane. If the center of the bore also lies in a vertical plane forming an angle of 41.95° with a true north line, any sighted person peering through the hole at 3:00 A.M. solar time on a summer solstice will see the center of the sun's surface.

In order to obtain the formula for the time of sunrise, the time when the altitude of the sun is 0°, again apply the ubiquitous cosine law. We need to discover what the value of γ is when $c = 0^\circ$. The preceding considerations generate the equation $\cos 90^\circ = 0 = \sin lat \sin dec + \cos lat \cos dec \cos \gamma$. Therefore, $\cos \gamma = -(\tan lat)(\tan dec)$, whence,
$\gamma = \text{Arccos}[-(\tan lat)(\tan dec)] = 180^\circ - \text{Arccos}[(\tan lat)(\tan dec)]$.

Since the sun's vertical rays proceed westward at the rate of 15 degrees per hour, the number of hours between solar noon and sunrise is $\frac{\gamma}{15^\circ}$. Thus the time of sunrise is $12 - \frac{\gamma}{15^\circ}$.
On substituting the previously determined value of γ, the magic of algebra condenses the expression and informs us that:

$$\textbf{Solar time of sunrise is: } \frac{\textbf{Arccos}[(\textbf{tan}\,lat)(\textbf{tan}\,dec)]}{\mathbf{15^\circ}}$$

From this we can also obtain the time of sunset, since solar noon is midway between sunrise and sunset.

If we now let *dec* equal its value at equinox, 0°, we may mathematically certify that on an equinox the solar time of sunrise is 6:00 A.M. Similarly, for an observer on the equator the 0° latitude provides a computed solar time 6 o'clock sunrise *every day of the year*.

Of course, a correction for the time by which solar noon at your longitude differs from clock noon must be made when computing the time of your sunrise. Even when solar noon and clock noon are considered identical, for a given time zone, they will be identical only on the central meridian of that time zone. The time of sunrise supplied by the formula will indicate the time of sunrise only at that central meridian. So a 4 minute time correction must be made for each degree difference between your longitude and that of your time zone's central meridian. The longitudes of these central meridians are always multiples of 15 degrees. For example, if your longitude is 80 degrees then the central meridian of your time zone (Eastern) must be 75 degrees. This positive 5 degree longitude difference will produce a 20 minute increase in your computed time of sunrise. Also, if our observation is not at sea, sunrise is not observed from the surface of a perfectly spherical earth and that introduces an error. In addition, atmospheric refraction of sunlight will slightly alter the observed time of sunrise.

Notice that when the sun rises at equinox, $dec = 0^{\circ}$, alt $= 0^{\circ}$ and $time\ angle = 90^{\circ}$. Therefore, applying the formula for azimuth, $\sin az = \dfrac{\sin(time\ angle)\cos dec}{\cos alt}$, we see that on an equinox all the trigonometric functions involved are equal to 1, including sin az. From this we infer that $az = 90^{\circ}$ and confirm the following much disseminated truth: At equinox the sun rises due east.

PROBLEM: We are given that the declination at the summer solstice is 23.45° north. Assume that the moment of solstice is precisely at the moment of sunrise, and then, for an observer at 40° north latitude, determine the direction of the rising sun at the summer solstice.

Using the given information we first determine the time of sunrise. This results from the equation:

$$\text{Time of sunrise} = \frac{Arc\cos(\tan 40^{0}\ \tan 23.45^{0})}{15^{0}}$$

With this we learn that sunrise occurs at 4.58 hours. (Avoid the temptation of reading this as 4:58.) Thus the number of hours from sunrise to solar noon is 7.42 hours. It follows that the *time angle* is 7.42 hours times 15° or 111.3°. Using

$$\sin az = \frac{\sin(\textit{time angle})\cos \textit{dec}}{\cos \textit{alt}},$$ we have

$$\sin az = \frac{\sin 111.3^\circ \cos 23.45^\circ}{\cos 0^\circ}.$$

On solving, $az = 58.7^\circ$, in which case the sun rises 31.3° north of east. By the time this rising sun sets, the declination will have decreased slightly, so the comparable angle at sunset will be slightly smaller. At the December solstice, the sunrise and sunset angles will be equivalent, but lie to the south.

An interesting aspect of sun induced shadows can be determined with the aid of the relations developed in this chapter. On an equinox, the tip of the shadow of a stationary object moves along a straight west-east line. The declination is 0° for only an instant of the day of equinox, but at no time during that day does the shadow deviate significantly from the west-east line. For example, if the declination is precisely 0° at noon, the tip of the shadow of a six foot pole positioned at 40° latitude would move approximately 29 feet between noon and 5 o'clock, but the tip of the shadow of the pole would stray from the west-east line by only 0.7 inches.

Since the sun's declination is so vital to our calculations, the procedure for determining it, defined by the following steps, is surely welcome. The values it provides are exceptionally close to those listed in an astronomical almanac.

1. Determine the number of days since the last solstice or equinox event. If the precise moment of the event is obtainable, so much the better.

2. Divide this number by the number of days from the initial equinox or solstice to the succeeding solstice or equinox.

3. Multiply this ratio by 90°.

4. If the initial event in the time frame is a solstice, multiply the cosine of the last result by 23.45°. The result will be the absolute value of the declination. If the initial event is an equinox instead of a solstice, then multiply by the sine instead of the cosine.

After the autumnal equinox but before the vernal equinox the sun will be south of the equator and the declination must be assigned a negative value.

If you are able to determine that time during the day when the length of the sun's shadow is minimum, that will be the time when the sun's altitude is maximum and this moment coincides with the time of solar noon.

To find an observer's longitude consult a solar almanac to determine the time of solar noon at the prime meridian (0° longitude). Then compute the time in hours between the prime meridian solar noon and the observer's solar noon. For an observer in the western hemisphere, the observer's longitude would then be obtained by multiplying that figure by 15°.

So with solar observations, and solar noon data, we are now able to calculate our longitude as well as our previously explored latitude. At long last the reader has the capability to determine whether or not that expensive global positioning system recently purchased is providing reliable coordinates.

SUMMARY OF SOLAR TERMS AND FORMULAS

TERMS

The symbol *alt* represents the sun's altitude or angle of elevation of sun.

The symbol *az* represents the sun's azimuth, the clock-wise angle measured from true north to a vertical plane through the sun.

The symbol *dec* represents the sun's declination, or latitude of the sun's vertical rays.

The symbol *lat* represents the latitude of the observer.

FORMULAS

1. *Time angle* $\gamma = 15^{\circ} \times$ (number of hours from time of solar observation to solar noon)
2. At solar noon, $alt = 90^{\circ} - lat + dec$.
3. Altitude of sun:
   $\sin alt = \sin lat \sin dec + \cos lat \cos dec \cos(time\ angle)$
4. Azimuth of sun: (a) $\sin az = \dfrac{\sin(time\ angle)\cos dec}{\cos alt}$
   (b) $\cos az = \dfrac{\sin dec - \sin lat \sin alt}{\cos lat \cos alt}$
5. Time of sunrise $= \dfrac{Arc\cos(\tan lat \tan dec)}{15^{o}}$, subject to corrections as stated on page 103.
6. Sun's azimuth at sunrise: (a) $\sin az = \cos dec \sin(time\ angle)$
   (b) $\cos az = \dfrac{\sin dec}{\cos lat}$

A Morsel (food for thought)

Not all branches of mathematics involve the concept of a number. Those branches that do, have determined that there are five numbers that are fundamental. Three of these, 0, 1 and π should be familiar to all. In addition there is the number e – Euler's Number – the base of natural logarithms, and the number $i = \sqrt{-1}$. This latter number is called an imaginary number since it can't be used for counting or measuring. Nevertheless it is indispensable, having many practical applications. Another of the most pervasive concepts in mathematics is that of the equation. Intriguingly these concepts all merge in the following statement.

$$e^{\pi i} + 1 = 0$$

For this we are indebted to the prolific Swiss mathematician and physicist, Leonhard Euler. 1707-83, authored the first calculus book.

CHAPTER 21

MUCH ADO ABOUT ORBITS

It will never be said of this chapter, "It doesn't take a Rocket Scientist." So strap in, check your instruments, listen to the countdown and get ready to lift off. We will embark on a comprehensive investigation of the motion of a satellite in orbit about a central body. The results will be based on the assumption that the only gravitational forces involved in our analyses are those between the two bodies mentioned.

From the polar coordinate equation of a satellite's path, provided later, it can be shown that the path will be either an ellipse, a circle, a parabola or an hyperbola. We will ignore the parabola and hyperbola because a satellite in such a path will never return. The remaining closed circular and elliptical paths are called orbits. Since a circular path creates little to investigate, we will concentrate on a study of the elliptical orbit.

DIAGRAM 1

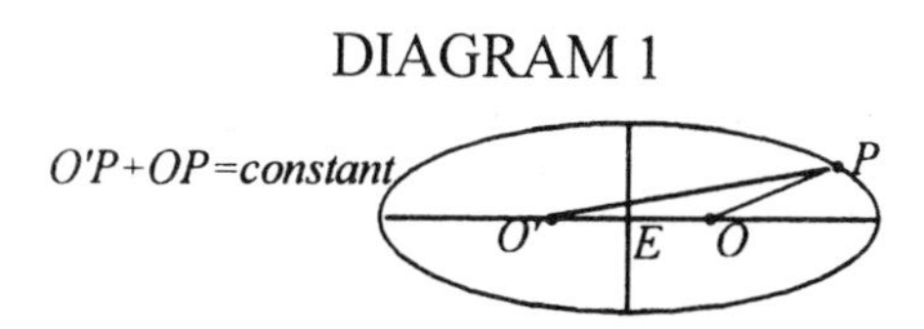

DEFINITIONS OF TERMS AND SYMBOLS

An ellipse (Diagram 1) is the set of all points P such that the sum of the distances from two fixed points, O and O', is a constant. Each of the fixed points is called a focus. The line segment passing through O and O' which lies within the ellipse is called the major axis. The minor axis, also within the ellipse, is the segment which perpendicularly bisects the major axis. The intersection of these two axes, E, is the center of the ellipse.

DIAGRAM 2

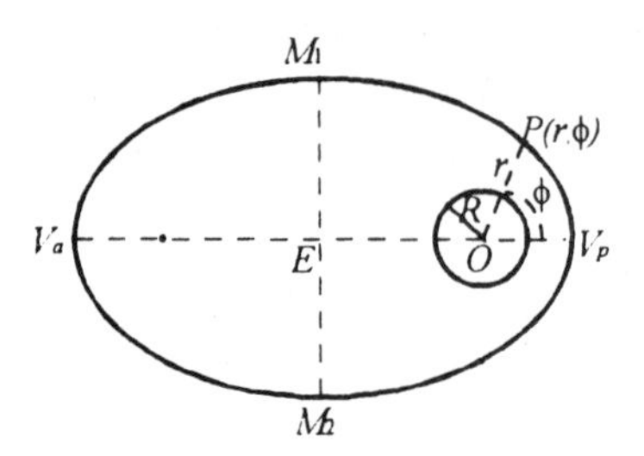

For our elliptical orbit we refer to Diagram 2. Point P is our position as we revolve about the central body. The center of the central body will always be at one of the foci. We have shown our central body at the focus O. Two points in the orbit of the satellite will be of primary interest to us–the point nearest to and the point farthest from the central body. If the central body is the sun, the nearest point is called the perihelion and the farthest point is called the aphelion. If the earth is the central body, these points are called perigee and apogee. V_p is the perihelion (or perigee) and V_a is the aphelion (or apogee). These points, called vertices, are the end points of the major axis. From the definition of ellipse, it can be shown that the sum of the distances from point P to the foci is always equal to the length of the major axis segment , V_aV_p. When we examine the polar coordinate equation for this orbital system we will choose the origin to be point O–the center of our central body.

The eccentricity of an ellipse, which will be represented by the letter e, is the ratio of distances, EO/EV_p. The value of the eccentricity, which is always a number between zero and 1, determines the shape of the ellipse. If the value is near zero, the ellipse is nearly circular, and if the value is near 1, the ellipse is long and narrow. The eccentricity of the earth's orbit is approximately 1/60.

Of particular interest will be man-made satellites which are inserted into orbit by means of rocket propulsion. In this case, we will learn how the satellite's orbit is defined by the rocket's direction and speed at time of power shut-off.

The symbols defined below will repeatedly appear in the derivations which follow. (Refer to Diagram 2).

A = length of semi-major axis = segment EV_p

B = length of semi-minor axis = segment EM_1

C = distance from center to a focus = segment EO

a = distance from point O to aphelion or apogee = segment OV_a

p = distance from point O to perihelion or perigee = segment OV_p

R = radius of central body

g = the acceleration of gravity at surface of central body

e = eccentricity of elliptical orbit = $\frac{C}{A}$

r = distance from point O to point P on orbiting satellite

$\phi = \angle POV_p$

r_0 = distance to satellite at power shut-off

v_0 = satellite speed at power shut-off

$\underline{A}$ = area of ellipse

s = distance traveled by the satellite

$v = \dfrac{ds}{dt}$ = speed of satellite

v_a = satellite speed at aphelion (apogee)

v_p = satellite speed at perihelion (perigee)

The values of a and p determine all characteristics of the ellipse and the satellite motion, so we will express all results as functions of a and p. For example:

$$A = \frac{a+p}{2}$$

$$C = \frac{a-p}{2}$$

If we refer to Diagram 2 and the right triangle with vertices at O, E and M_1, it can be shown that

$$B^2 = A^2 - C^2. \text{ Hence } B^2 = \frac{(a+p)^2}{4} - \frac{(a-p)^2}{4} = ap.$$

Therefore

$$B = \sqrt{ap}.$$

Interestingly then, A = the arithmetic mean of a and p, while B = the geometric mean of a and p. The eccentricity mentioned previously is defined as the ratio C/A, which may be transformed as follows:

$$e = \frac{C}{A} = \frac{(a-p)}{(a+p)}$$

Since the area of the ellipse is πAB we have

$$\underline{A} = \frac{\pi}{2}(a+p)\sqrt{ap}$$

In order to proceed further, it will be necessary to refer to the polar coordinate equation of the satellite's path. We assume that a satellite is being inserted into an orbit about the earth. We also assume that at the instant of power shut-off, the satellite is at point V_p whose polar coordinates are $(r_0,0)$. In addition, it is moving perpendicularly to the major axis in a counterclockwise direction and is traveling at a speed of v_0 at time of power shut-off. The example contemplated in the design of Diagram 2 implies that $r_0 = p$. Whenever the power shut-off conditions just mentioned are satisfied, r_0 is equal to either p or a. If we build on Newton's revelation that the gravitational attraction between any two bodies is inversely proportional to the square of the distance between their centers of mass, it is possible, with sufficient analysis, to arrive at the following finding. *The polar coordinate equation of the path of a satellite is:*

$$r = \frac{r_0^2 v_0^2}{gR^2} \div \left[1 + \left(\frac{r_0 v_0^2}{gR^2} - 1 \right) \cos\phi \right]$$

It can be shown that the absolute value of the coefficient of $\cos\phi$ is the eccentricity, that is,

$$e = \left| \frac{r_0 v_0^2}{gR^2} - 1 \right|$$

From analytical geometry, we learn that the path will be a circle if the eccentricity is equal to zero, an ellipse if the eccentricity is between zero and one, a parabola if the eccentricity is equal to one, and an hyperbola if the eccentricity is greater than one.

Thus, we know that: If $\frac{r_0 v_0^2}{gR^2}$ <1, the path will be an ellipse and the point of power shut-off becomes the apogee of the orbit. If $\frac{r_0 v_0^2}{gR^2}$ = 1 the orbit will be the circle $r = r_0$. If 1<$\frac{r_0 v_0^2}{gR^2}$<2, the path will be an ellipse with point of power shut-off now the perigee. If $\frac{r_0 v_0^2}{gR^2}$ = 2, the path will be a parabola. In this case the satellite will escape the earth's gravity. Finally, if $\frac{r_0 v_0^2}{gR^2}$>2, the path will be an hyperbola, and the satellite will again escape earth's gravity. Remember that this point of power shut-off will be the perigee or apogee only if the position vector from the earth's center to the satellite is perpendicular to the satellite's path at time of power shut-off. In the process of deriving the polar equation of a satellite in orbit (see preceding page), the author ascertained that the time for one revolution of a satellite about the central body is:

$$T = \pi \frac{(a+p)^{1.5}}{R\sqrt{2g}}$$

This result is consistent with one of Kepler's laws, which states that the period of revolution of a planet varies as the 1.5 power of its distance from the sun.

Here the readers are given two problems which may be of interest. It is suggested that distances be measured in feet, and time in seconds.

1. The mean distance to a satellite is often a good approximation to the average of the distances to apogee and perigee. The mean distance of the moon from earth is 238,857 miles. So $a + p$ would be twice this distance. Using 3960 miles as the earth's radius and $g = 32.15$ ft/sec^2, determine the time for one revolution of the moon about the earth. Your answer should lie between 27 and 28 days.

2. Using 365.24 days as the length of the earth's year, 91.5 million miles as the earth's distance from sun at perihelion and 94.5 million miles as distance at aphelion, radius of the sun 432 thousand miles, determine the acceleration of gravity at the surface of the sun. Your answer should be in the neighborhood of 900 ft/sec^2. Remember that the effect of other bodies in the solar system, which our formulas ignore, will slightly affect answers.

Another of Kepler's laws is: A position vector from center of central body to center of satellite will sweep out the area of the ellipse at a constant rate. This constant rate, $\frac{dA}{dt}$, in area per second, may be found by dividing the area of the orbital ellipse by the time it takes to sweep out the ellipse's entire area. Since this latter value is the period of revolution, mentioned previously,

$$\frac{dA}{dt} = \left[\frac{\pi}{2}(a+p)\sqrt{ap}\right] \div \left[\frac{\pi(a+b)^{1.5}}{R\sqrt{2g}}\right].$$

So we have the important formula:

$$\frac{dA}{dt} = R\sqrt{\frac{gap}{2(a+p)}}$$

Now we borrow two formulas from the calculus. The rate at which the area of the ellipse is swept out is given by the formula

$$\frac{dA}{dt} = \frac{1}{2} r^2 \frac{d\phi}{dt}$$

and the speed of the satellite is

$$\frac{ds}{dt} = \sqrt{\left(\frac{dr}{dt}\right)^2 + r^2 \left(\frac{d\phi}{dt}\right)^2}$$

With the aid of these two formulas we are able to determine the speed of the satellite at perigee and apogee, namely, v_p and v_a. Since r is a minimum at perigee and a maximum at apogee, $\frac{dr}{dt} = 0$ at both points. Therefore, the formula for the speed of the satellite at perigee and apogee reduces to $\frac{ds}{dt} = r\frac{d\phi}{dt}$.

Consequently, $v_p = p\frac{d\phi}{dt}$ and $v_a = a\frac{d\phi}{dt}$.

So, at perigee, $\frac{dA}{dt} = \frac{p}{2} v_p$ and at apogee $\frac{dA}{dt} = \frac{a}{2} v_a$. Equating each of these to the previously determined $\frac{dA}{dt}$, we see that:

$$v_p = R\sqrt{\frac{2ga}{p(a+p)}}$$

$$v_a = R\sqrt{\frac{2gp}{a(a+p)}}$$

It can be shown that the r_0 in the equation of the elliptical orbit is equal to a when $r_0 v_0^2 < gR^2$, but otherwise $r_0 = p$. In either case

$$(r_0 v_0)^2 = \frac{2gR^2 ap}{(a+p)}.$$

With this result and the knowledge that the absolute value of the coefficient of $\cos\phi$ is also equal to the eccentricity of the ellipse, $e = \dfrac{a-p}{a+p}$, we can transform the formula for r, on page 113 to:

$$r = \frac{\left[\dfrac{2ap}{a+p}\right]}{\left[1 \pm \dfrac{(a-p)}{(a+p)}\cos\phi\right]}.$$

The minus sign is chosen if and only if $r_0 v_0^2 < gR^2$. If we multiply numerator and denominator by $\dfrac{a+p}{2}$, the latter form of the polar coordinate equation simplifies to:

$$r = \frac{B^2}{A \pm C\cos\phi},$$

where A, B and C are respectively the previously mentioned semi-major axis, semi-minor axis and distance from center to a focus.

From this relation we can obtain the formula for $\dfrac{dr}{dt}$, enabling us to develop our final result, the expression which represents $v = \dfrac{ds}{dt}$, where v is the variable speed of the satellite as a function of ϕ. We first note that the aforementioned $\dfrac{dr}{dt}$ is given by:

$$\frac{dr}{dt} = \mp \frac{B^2 C\sin\phi}{\left(A \pm C\cos\phi\right)^2}\frac{d\phi}{dt}.$$

Using the similarities between the formulas for r and $\frac{dr}{dt}$, we see that,

$$\frac{dr}{dt} = \mp \frac{Cr^2 \sin\phi}{B^2} \frac{d\phi}{dt}$$

Replacing B and C by their known functions of a and p, we have

$$\frac{dr}{dt} = \mp \frac{(a-p)}{2ap} r^2 \sin\phi \frac{d\phi}{dt}$$

Next, we again use

$$\frac{dA}{dt} = \frac{r^2}{2}\frac{d\phi}{dt} = R\sqrt{\frac{gap}{2(a+p)}}$$

from which we obtain

$$r\frac{d\phi}{dt} = \frac{R}{r}\sqrt{\frac{2gap}{a+p}}.$$

We are now able to find v, our final formula.

Beginning with the calculus derived formula for $\left(\frac{ds}{dt}\right)^2$, we have

$$v^2 = \left(\frac{ds}{dt}\right)^2$$

$$= \left(\frac{dr}{dt}\right)^2 + \left(r\frac{d\phi}{dt}\right)^2$$

$$= \frac{gR^2}{2ap(a+p)r^2}\left\{\left[(r\sin\phi)(a-p)\right]^2 + (2ap)^2\right\}$$

Notice that if the path is a circle, $r = a = p = r_0$, and the speed reduces to $R\sqrt{\frac{g}{r_0}}$. This is consistent with the previous statement that the path is a circle when $\frac{r_0 v_0^2}{gR^2} = 1$. If $\phi = 0^\circ$ or 180°, the satellite is at apogee or perigee. Again our formula for v is consistent, since it then reduces to the previously determined v_a or v_p.

When in elementary physics we study the path of a projectile in a vacuum, it is assumed that the gravitational force vectors are always perpendicular to a horizontal ground. This implies that the gravitational vectors at any two positions are parallel. This assumption dictates that the path be parabolic. However, the assumption distorts the facts since it ignores the principle that the gravitational force vectors are all directed toward the center of the earth and hence are not quite parallel. As a result of the preceding discourse we now know that the path will always be an ellipse instead of a parabola. The elementary approach used is understandable, since it makes the description of the motion much more tractable and is sufficiently accurate for most elementary work. For a comparison of the two approaches we consider the following problem.

A rifle bullet with a muzzle velocity of 3000 feet per second is propelled horizontally from a point 1000 feet above a smooth lake. Where does the bullet strike the water? Allow yourself the luxury of an environment uncluttered by atmosphere.

DIAGRAM 3

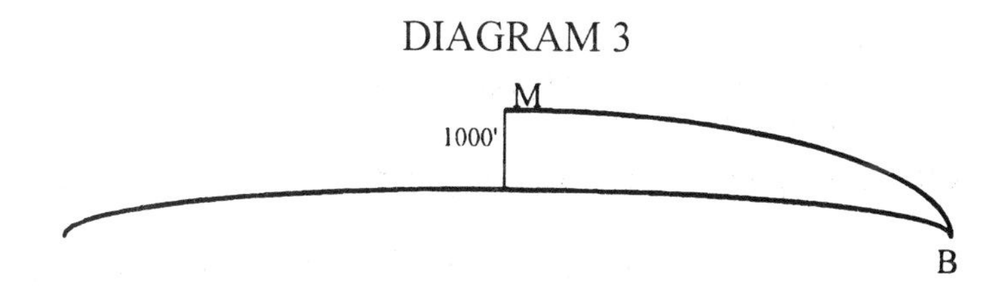

To produce the solution, we must first obtain the polar equation of the bullet's elliptical path. Replace the general values in the polar coordinate equation given on page 113 with our problem's specific values. In order to obtain this polar equation directly, the segment joining the center of the earth and the gun muzzle must be chosen as the 0 degree axis. We then use the following information: The muzzle position is the apogee, $v_0 = 3000$ ft/sec, and $a = r_0 = R + 1000$ ft. Also we know that $R = 3960$ miles and $g = 32.15$ ft/sec^2.

The value of the eccentricity e turns out to be 0.9866. Since $e < 1$, the elliptical path is confirmed, and its proximity to 1 confirms that the path is almost a parabola. Since the point at which the bullet strikes is the intersection of its path with the surface of the lake, we need only to find the ϕ-coordinate of the ellipse for which the r-coordinate is R. Then multiplying the radian measure of ϕ by the radius of the earth, R, provides the distance to the point where the bullet strikes the water. The solution turns out to be 23,822 feet, which is 160 feet greater than the solution provided by the methods of elementary physics.

An interesting fact about projectiles is provided by one of our formulas, namely,

$$\frac{dA}{dt} = \frac{1}{2} r^2 \frac{d\phi}{dt}.$$

We have stated that this product remains constant. Suppose an object is propelled vertically, in a vacuum, from point P on the earth's equator. Since the earth rotates from west to east, its west-east velocity provides a component which helps propel the object into an elliptical orbit as shown in Diagram 4 on the next page.

DIAGRAM 4

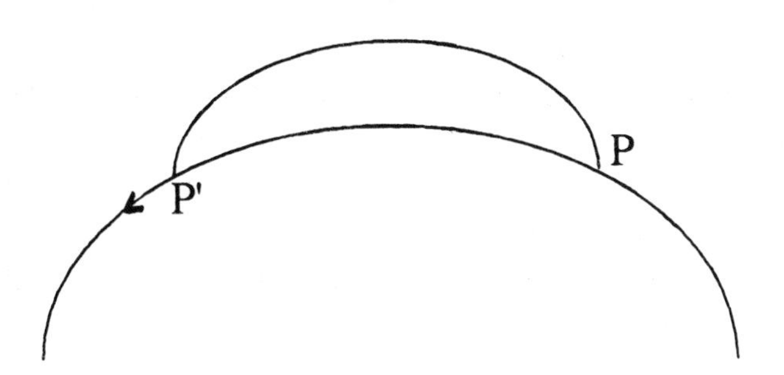

When the object is on its elliptical path its r-coordinate is greater than it was when it left point P on the equator. So, during flight, $\frac{d\phi}{dt}$ for the object must then be less than $\frac{d\phi}{dt}$ for point P. This is because $r^2\frac{d\phi}{dt}$ remains constant and the value of r for the object is greater than the value of r for point P. As a result, during flight, the ϕ-coordinate of the object changes less than the ϕ-coordinate of point P on the equator. This means that an object propelled vertically from the earth's equator will land slightly to the west of point P', the rotated position of the point P from which the object was launched. The problem is more complex if the object is propelled from a point not on the equator. The object's elliptical path will, as always, lie in a plane which passes through the center of the earth and is perpendicular to the meridian from which the object is vertically propelled. The object must then land on the great circle of intersection of this plane with the earth's surface. Since the point P from which the object departs is the point of this great circle most distant from the equator, the object will land to the southwest of a point P in the northern hemisphere, but to the northwest of a point P in the southern hemisphere. With the aid of the formulas we have developed, it is possible to determine the point at which the object will land, but the solution is tedious.

A summary of the preceding formulas describing orbits and orbital motion now follows,

Relevant symbols are:

a = distance to aphelion or apogee

p = distance to perihelion or perigee

R = radius of central body

g = acceleration due to gravity at surface of central body

The numbers in brackets indicate some of the corresponding computed values as exemplified by the earth's orbit. For the earth's orbit $a = 94.512$ million miles, $p = 91.402$ million miles, $R = 432{,}000$ miles, $g = 900.3$ ft/sec^2.

FORMULA SUMMARY

1. $A = \dfrac{(a+p)}{2}$, semi-major axis of ellipse

2. $B = \sqrt{ap}$, semi-minor axis of ellipse

3. $C = \dfrac{(a-p)}{2}$, distance center to focus of ellipse

4. $\underline{A} = \dfrac{\pi}{2}(a+p)\sqrt{ap}$, area of ellipse

5. $e = \dfrac{(a-p)}{(a+p)} = \dfrac{C}{A} = \left|\dfrac{r_0 v_0{}^2}{gR^2} - 1\right|$, eccentricity of ellipse, [0.0161]

6. $T = \dfrac{\pi(a+b)^{1.5}}{R\sqrt{2g}}$, time for one revolution of orbit, [365.37 days].*

7. $v_a = R\sqrt{\dfrac{2gp}{a(a+p)}}$, speed at apogee, [96,100 ft/sec]

8. $v_p = R\sqrt{\dfrac{2ga}{p(a+p)}}$, speed at perigee, [99,200 ft/sec]

9. $av_a = pv_p = R\sqrt{\dfrac{2gap}{(a+p)}} = 2\dfrac{dA}{dt} = r^2\dfrac{d\phi}{dt}$

10. $r = \dfrac{r_0^2 v_0^2}{gR^2} \div \left[1 + \left(\dfrac{r_0 v_0^2}{gR^2} - 1\right)\cos\phi\right]$, polar equation of orbit

11. $r = \dfrac{B^2}{A \pm C\cos\phi}$, minus sign if $r_0 v_0{}^2 < gR^2$, alternate form of polar equation.

12. $v^2 = \dfrac{gR^2}{2ap(a+p)r^2}\left\{\left[(r\sin\phi)(a-p)\right]^2 + (2ap)^2\right\}$, square of satellite speed for any ϕ.

* This slight inaccuracy is mainly due to the lack of accuracy in the given measurements for R and g.

A Morsel (food for thought)

When I was a boy, temperatures were measured in degrees Fahrenheit or Centigrade. Then in 1948 some eggheads got together and told us that we must now call the Centigrade scale the Celsius scale. No matter what we call it, the next formula indicates how the Fahrenheit temperature and the other one are related. The symbol F represents Fahrenheit degrees and C represents degrees Celsius.

$$F = 1.8C + 32$$

This enables us to solve the following simple algebra problem. At what temperature is the Fahrenheit temperature equal to the Celsius temperature?

The answer is the year in which Calvin Coolidge was elected, president subtracted from the year in which Grover Cleveland was elected to his first term.

CHAPTER 22

FOUR FOR THE ROAD

On the way out, we leave the readers with the following four problems. Problems 2, 3 and 4 require a knowledge of elementary calculus and physics. All four problems possess a uniqueness that has made them, for me, lastingly intriguing. The second one was assigned, as an extra credit problem to most of my calculus classes at the Colorado School of Mines. During that eighteen year period, only one student solved it. So a salute goes to Larry Webster. Solutions or information about solutions follow the statement of the problems.

Statement of the Problems

1. Initially, each door of a long line of lockers, numbered in order and beginning with number 1, is closed. A man goes down the line of lockers and opens each locker door. Then a woman goes down the line and closes every second door. Next a person goes to every third door, closing each door that is open and opening each door that is closed. Following this, an individual proceeds to every fourth door and closes each door that is open and opens each door that is closed. This process is continued, such that the nth person changes the position of the doors on every locker whose number is a multiple of n. The process ends when the first door moved is also the last door in line. Describe the class of numbers on the doors which will be open at the end of the process.

2. A snow plow started plowing at noon. It traveled twice as far during the first hour as it did during the second hour. What time did it start to snow?

3. Farmer Brown has a truck with square wheels. This is fine with Farmer Brown since these wheels allow his truck to travel along his road with absolutely no bouncing. What is the shape of the surface of his road?

4. There is a function $f(t)$ for which $\int_{x}^{x^2} f(t)dt \equiv 1$. First graph the function and then determine the $f(t)$.

About the Solutions

1. If the door has been moved an odd number of times, it will be open at the end of the process. If the door has been moved an even number of times, it will be closed at the end. A door will be moved one time for each factor of its number. Hence the door will be open at the end if and only if its number has an odd number of factors. If an integer is not a square, its factors occur in pairs, one smaller than the square root of the number and the other greater than the square root. For example, the integer 12 has three pairs of factors 1 and 12, 2 and 6, 3 and 4. If an integer is a perfect square, such as 16, it has pairs of factors 1 and 16, 2 and 8, but it also has an unpaired factor, namely, its square root, 4. So the number of factors of an integer will be odd if and only if its square root is an integer. Thus a locker will be open at the finish if and only if its number is a perfect square.

2. The problem's proposer expected the solver to make the reasonable assumptions which are necessary to construct a sensible problem – such as the snow continues to fall at a constant rate and the plow produces constant power. In this case, it must have started to snow at 11:23 A.M. To obtain this result:
   let w = work, s = distance traveled, F = force, P = power, t = time of travel. In this case, dw = Pdt = Fds = (ct)ds.

3. Every vertical cross section parallel to the center line of the road would be a series of identical inverted arcs of catenaries. The length of each arc is equal to the length of a side of a wheel, and the pair of tangents at each cusp forms a right angle.

4. For large values of x there will be a large interval from x to x^2. Therefore, since the value of the integral is constant, the graph must be asymptotic to the positive t axis. If x is infinitesimally larger than 1, the interval from x *to* x^2 will also be infinitesimal. As a consequence, the graph must be positively asymptotic to the vertical line $t = 1$. This latter reasoning also applies for x slightly larger than 0 or slightly smaller than 1. But for $0 < x < 1$, the value of $x^2 < x$. This means that the dt factor will be negative and, in order for the integral of $f(t)dt$ to be positive, the factor $f(t)$ must also be negative. Hence, for $0 < x < 1$, the graph is negatively asymptotic to the vertical lines $t = 0$ and $t = 1$.

   To determine $f(t)$, convert the integral whose limits are x *to* x^2 to the difference of two integrals, with limits k to x^2 on the first integral and limits from k to x on the second integral. Equating this difference of integrals to 1 provides the necessary alternate form for the integral equation. Now we equate the derivatives, with respect to x, of both sides of the integral equation. Recall that $\frac{d}{du}\int_k^u f(t)dt = f(u)$. Applying this principle to the preceding integral, will lead to the relation $\frac{f(x)}{f(x^2)} = 2x$. Sufficient ingenuity and common sense may supply $f(t) = \frac{1}{(\ln 2)t\ln t}$.

I don't know what I'll do with myself,
now that Winton's book is finished.

GLOSSARY

Aliquot part - the decimal equivalent of a simple fraction.

Angstrom - 10^{-8} centimeters.

Aphelion - the point on the orbit of a satellite orbiting the sun which is farthest from the sun.

Apogee - the point on the orbit of a satellite orbiting the earth which is the farthest from the earth.

Asymtote - a line which is continually approached by a curve as the line and curve are infinitely extended.

Azimuth - the angular distance measured clockwise from either true north or true south on the horizon circle, to that point on the horizon circle which is directly below or above the line of sight to the observed object.

Binomial - a mathematical expression consisting of 2 terms connected by a plus or minus sign.

Binomial Theorem - the theorem which describes the set of terms generated when a binomial is raised to any specified power.

Catenary - the curve formed by a perfectly flexible cord or chain of uniform density, hanging freely between 2 points at the same height.

Coefficient - a number or algebraic symbol prefixed as a multiplier to a variable or unknown quantity as in 3x and ax. The 3 and a are the coefficients of x.

Ellipse - the set of all points in a plane such that the sum of the distances from 2 fixed points is a constant.

Equinox - the precise time when the sun's vertical rays cross the equator. On these occasions, day and night are of equal length everywhere.

Geometric Sequence - any sequence of numbers for which the ratio between consecutive terms remains constant. For example: 3, 6, 12, 24, 48, ...

Golden Ratio - that value of the ratio m/n for which $m/n = (m+n)/m$.

GLOSSARY

Golden Rectangle - any rectangle for which the ratio of length to width is equal to the Golden Ratio.

Great Circle - a circle on a sphere whose plane intersects the center of the sphere.

Hyperbola - the set of all points in a plane such that the difference in the distance from two fixed points is a constant.

Involute - the curve traced by the end of a taut string when it is wound around or unwound from a fixed plane curve.

Latitude - the angular distance, normally measured in degrees, from a point on earth to the earth's equator.

Linear Expression - $ax + b$, where a and b are constants, $a \neq 0$. The graph of the corresponding function is a straight line.

Longitude - the angle between the plane of the meridian through a given point and the plane of the prime meridian.

Meridian - the half of earth's great circle which passes through the geographical poles and any given point on earth. It is the arc of equal longitude.

Parabola - the set of all points in a plane equidistant from a fixed line and a fixed point not on the line.

Parallels of latitude - small circles on earth parallel to the equator.

Perigee - the point on the orbit of a satellite orbiting the earth which is nearest the earth.

Perihelion - the point on the orbit of a satellite orbiting the sun which is nearest the sun.

Prime or 0° Meridian - the meridian passing through Greenwich, England.

Prime number - a number which has only itself and the number one as factors.

Quadratic Expression - $ax^2 + bx + c$, where a, b and c are constants, $a \neq 0$.

GLOSSARY

Radian - the angle subtended by a circular arc whose length equals the radius. 1 rad = $180°/\pi$.

Solstice - the precise times when the sun's vertical rays are at the northernmost position (tropic of Cancer) or the southernmost position (tropic of Capricorn).

Tachometer - an instrument used to determine speed, especially the rotational speed of a shaft.

Numerical Cryptogram

If we replace each of the different letters with one of the digits zero through 9, the result will be a valid addition problem. Determine the relationship between letters and digits.

```
  S E N D
  M O R E
---------
M O N E Y
```

INDEX OF FORMULAS

INDEX OF FORMULAS

AFTERWORD

Thank you for sharing with me the joys of this endeavor. Also be it known that I am pleased with the money you contributed toward the defrayment of expenses. May you feel that your funds have been well spent.

Direct questions, complaints, suggestions and glowing words of praise to me at this address.

Aftermath Publishing
1490 Rogers Street
Golden, Colorado 80401

1-800-501-3848
FAX (303) 279-3075

Winton Laubach

A MORSEL (food for thought)

Verbal Cryptogram

(of 16 two-letter words)

Hopefully the preceding morsel has whetted the reader's appetite for solving cryptograms. If so, the following verbal cryptogram may be a fitting dessert.

TX DG DS DT SX KH, DS DT

RW SX ZH XU RT SX IX DS.